The Planning of Iron and Steelworks

By

Friedrich A. K. Lüth   Horst König

Third completely revised and enlarged edition

Translated from the German by

Gordon Cockburn

With 60 figures and tables

Springer-Verlag Berlin Heidelberg GmbH

1967

Dr.-Ing. FRIEDRICH A. K. LÜTH
Privatdozent, University of Aachen,
Consulting Engineer

Dr.-Ing. HORST KÖNIG
Director, Planning Division, DEMAG A. G.

GORDON COCKBURN
F.I.L., BDÜ, Member of the Translators' Guild

The first and second editions of this book were published in German
by these one publisher ublisher under the title:

Plainning und Bau von Hüttenwerken"

Additional material to this book can be downloaded from http://extras.springer. com

ISBN 978-3-662-28160-4   ISBN 978-3-662-29671-4 (eBook)
DOI 10.1007/978-3-662-29671-4

Copyright 1952 Springer-Verlag Berlin Heidelberg
© by Springer-Verlag Berlin Heidelberg 1967
Originally published by Springer-Verlag, Berlin/ Gottingen/ Heidelberg in 1967
Library of Congress Catalog Card Number 67-28563
Title-No. 1447

Preface to the Third Edition

At the present time, we are witnessing a period of vigorous growth in the iron and steel industries of the older industrialised countries. Looking farther afield, we find that iron and steel industries are also being established in countries that have in recent years gained their independence, countries that hitherto had no iron and steelworks at all, or none of any great significance.

This situation produces a certain diversity in the planning and building of iron and steelworks. In the first case, high-capacity iron and steelworks are built on the basis of conventional, tried-and-tested processes, or existing plant is extended; in the second case, considerably smaller works have to be planned, and very often newer or unorthodox metallurgical processes must be used to cater for the specific raw materials available to the country in question.

Following this change in the tasks of the planning engineer, and in view of the new iron and steelmaking methods developed in the meantime, the decision was taken to revise and enlarge F. Lüth's "Planung und Bau von Hüttenwerken", the first and second editions of which were published in the German language by the Springer-Verlag in 1955 and 1957 respectively.

In addition, the services of a second well-known expert in this field were secured for the preparation of the revised edition.

This edition is based on the latest stage of development in the fields of steelworks engineering and metallurgy; it sets out the basic factors to be taken into consideration when planning iron and steelworks, together with the sequence in which the planning stages must be tackled.

The authors, both of whom have many years of practical experience in the operating and planning of iron and steelworks all over the world, hope that this book will be of assistance not only to the expert in the older industrialised countries, but also to specialists in other countries who are concerned with the planning of iron and steelworks. They also hope that careful study of the various factors covered will help to avoid mistakes being made when planning such works, mistakes that invariably have far-reaching consequences.

The great interest displayed in the subject of this book in industrial and developing countries alike prompted the publishers to have the 3rd Edition translated into the English language.

Düsseldorf/Duisburg, July 1967

Friedrich Lüth Horst König

Contents

Contents

Folding chart *(between pages 104 and 105)*
Examples of various tube manufacturing installations

Introductory Notes

Even though many countries of the world have not yet adopted the metric system, it was for a number of reasons thought that this system ought to be used throughout this book. Conversion tables and factors can be found on pages 174—177, and the abbreviations used are listed on page 178.

The investment and production costs quoted in the various chapters are based on West European conditions. The term "DM" is to be understood as meaning Deutschmarks of the Federal Republic of Germany, and the term "$" refers to USA dollars. The DM values were simply converted into $ for the convenience of readers more accustomed to thinking in terms of that currency. They must be adjusted from case to case to cater for any difference between the wage rates, etc. of Western Europe and those of the USA, for example.

The object of this edition of "The Planning of Iron and Steelworks" is to assist in the solving of planning problems by providing the interested parties with general technical information and data accumulated during many years of practical experience in the planning of such works.

As this book is mainly concerned with the *planning* side of iron and steelworks, only the *principal* features of the technical processes and plant, etc. are discussed; readers wishing further information will find a list of recommended literature on pages 179—184.

A. General Conditions

The reasons for building metallurgical plant can be many and varied. For example, a genuine market demand can prompt the establishment of the first iron and steelworks, the extension of existing plant, or the building of further new works. Again, the availability of extensive raw material deposits, e.g. ores and coal, or of great quantities of electric power can provide the initial impulse. It would at this early juncture be expedient to mention the fact that many countries which in earlier times could not entertain the idea of building metallurgical plant because the available heating and reducing coal could not be used in conventional processes, e.g. for blast furnaces, can now realise their wishes by exploiting new processes. Finally, the need to save foreign currency can make the domestic manufacture of iron and steel products and urgent necessity.

It must be stressed from the very outset that the erection of metallurgical plant can never be regarded as a final stage; experience has shown—and this applies to every country—that the establishment of new iron and steel plant must be followed by the establishment of a considerable number of other branches of industry if the profitability of the iron and steel industry is not to be jeopardised.

Iron and steelworks can be linked in various ways with other industrial plant such as cement works, power stations, fertiliser factories, and so on. However, the most important condition is that a further processing industry of adequate capacity is established in the vicinity of the iron and steelworks.

Further, numerous examples demonstrate a fact that at sight is rather astonishing; in countries which had no domestic iron and steel industry and which imported a certain quantity of finished products, i.e. rolled products, the amount imported temporarily fell in some cases after an iron and steel industry had been established and taken up production, yet after a certain time the imports rose to the old figure, and often even exceeded it. The reason for this is to be found in the fact that the establishment of an iron and steel industry so invigorated the entire national economy that steel consumption increased sharply.

It must be stressed that the actual planning work must be preceded by the clarification of a number of basic questions and the execution of preliminary investigations. This is of paramount importance and calls for the most scrupulous care and attention. Even though certain points such as national economy considerations,

situation with regard to raw material sources,
other operating media sources, such as those for gas, electric power, mineral oil, etc.,
transport situation,
labour market,
and the profitability estimate

are given special mention in this chapter, this does not mean that these aspects represent an exhaustive list; in fact, a whole range of additional and entirely different points can arise. However, these are not so decisive, and a detailed discussion of them would exceed the scope of this book.

I. National Economic Conditions

In many cases, the planning of a works for the production of pig iron, crude steel, and rolling mill products will in the main be prompted by the need to compensate for a deficit in such items in the country concerned. The requirement may arise from the need to increase exports, or to satisfy an increased domestic demand for consumer goods. Trade policy may also call for the domestic production of steel to meet the demand, instead of importing the amount needed to fill the gap. Moreover, the past ten years have shown that efforts made in numerous countries without iron and steel industries to attain self-sufficiency have exerted an extremely powerful influence on the establishment of domestic iron and steel industries. In most cases, these countries had adequate domestic deposits of the main raw materials, iron ore and coal, or at least one of these two.

In the case of countries with highly-developed industries but insufficient indigenous raw materials, the correct procedure will always be to import raw materials for the iron and steel industry and export finished products. In the table given below, we take

as an example a small country in the Far East and compare the
volume ratios obtaining between 1950—1962:

Total requirement of iron and steel products, and coverage of the same
(in 1,000 of tons)

Year	Total requirement	Domestic production	Imports	Exports
1950	49.4	14.3	35.1	—
1952	79.8	22.7	57.1	—
1954	137.9	65.9	72.0	—
1956	162.2	92.5	70.9	1.2
1958	197.7	131.4	78.0	11.7
1960	293.7	231.3	111.6	49.2
1962	277.1	221.4	112.7	57.0

This example, based on the developments in one small country,
clearly demonstrates the tendency towards an increase in the consumption of steel that is to be noted all over the world. In spite
of a significant increase in the domestic production of steel, imports
also increased to almost the same degree. The increase in the consumption of steel can in part be explained by the fact that the
domestic production of iron and steel products permitted the
growth of further-processing industries, which in turn permitted the
export of the processed iron and steel products on an increasing
scale.

Countries such as Turkey, Egypt, Brazil, and the Argentine
offer similar examples. In these countries, integrated iron and
steelworks which at the time of erection were even considered as
being too big for the existing demand were the subject of extension
planning work shortly after commissioning; in some cases, new
works had to be planned.

Even countries that have no significant raw material deposits are
making efforts to establish their own iron and steel and further-
processing industries. This is true of Japan, for example. Also,
countries that have extensive oil and natural gas deposits but no
ores or coking coal are now making efforts to use these inexpensive
raw materials for smelting ores, with the object of producing their
own steel.

The market conditions and domestic demand must be thoroughly
studied before the planning of a metallurgical plant for a non-
industrialised country is undertaken. Unfortunately, the import

statistics used as a basis often leave much to be desired. This work must also take into consideration the individual products such as heavy plate, medium plate, sheet, heavy sections, medium sections, light sections, and tubes, etc.

As it is uneconomical to build rolling mills for all these products, the rolling programme and the iron and steelwork's production programme must be arranged to cater for the rolled steel products and grades expected to be predominant in the future demand.

II. Plant Location and Raw Material Deposits

Formerly, the rule was that the site of every iron works was governed by the location of the two most important raw material deposits, iron ore and coal. This is no longer true in every case, as we have at our disposal new iron ore smelting processes which can use not only coal, but oil, natural gas, or inexpensive electric power, for example. The traditional industrialised countries with their big production units still have ore and coke as their basis, however, and this will remain so in the immediate future. But even countries with highly-developed industries that had no coking coal at their disposal made early use of other smelting processes, the main feature being the replacement of coking coal by electric power; coking coal need not always be the required carbon bearer. This process, the electric reduction process, has been introduced into South America, the Northern countries, and Italy to a particularly marked extent.

Examples of the establishment of iron and steel industries "on the ore" are to be found in the Siegerland, Alsace Lorraine, Salzgitter, Luxembourg, Krivoi Rog, Magnitogorsk, Styria, Corby (Northants.), Alabama and so on. Normally, the rule is that the iron and steelworks are erected on the coal when high Fe-content ores form the bulk, and on the ore when ores low in Fe are to be smelted, the main governing factor when classical processes are used being the transportation costs, of course. A remarkable and unique undertaking in the Soviet Union, an undertaking that will scarcely be duplicated anywhere in the world in the foreseeable future, was commenced about forty years ago; this is the linking of metallurgical plants located in the one instance on the coal and in the other on the ore. The iron industry area of Magnitogorsk in the Southern Urals, which has its own iron ore, is linked to the new industrial centre in the Kusnezk coal-mining area by a "Magis-tral", a four-track railway line 2,500 kilometres in length. High-capacity trains carrying 2,000 tons run non-stop between the two locations, taking ore in the one direction and coal in the other.

Metallurgical plants located away from both iron ore and coal deposits are usually to be found on the coast; examples of this are found at Ijmuiden, Bremen, Lübeck, Stettin, Mondeville, Genoa, Oxelösund, in England, and the coasts of the North American lakes. These coastal locations are made the most favourable by the availability of low sea-freight rates, and the possibility of smelting a variety of ores from different countries, taking full advantage of the world market situation obtaining at any particular time. The same advantage is also enjoyed, with certain limitations, by those inland works that have good inland waterway connections to big seaports; they, too, can take advantage of world-market prices. The Ruhr district and many British iron and steelworks can be in-cluded in this category. The trend towards erecting iron and steel-works on the sea-coast has undergone a remarkable intensification during the past few years. The proportion of coastal iron and steel-works in the member countries of the ECSC was only 6.6% in 1955; this figure rose to 12% as early as 1961, and by 1970 the percentage will be substantially higher. Special reference is made in this context to the high-capacity iron and steelworks erected along the coast of Italy: the coastal sites were deliberately chosen in view of the need to import iron ore and coal, and also in view of the favourable freight rates for exports, of course. The planning of new iron and steelworks on the coasts of Belgium, France, Holland, and Spain follows the same trend.

In many cases, investigations will have to be carried out to determine whether the smelting of indigenous ores with low Fe contents or the importation of high-grade ores is the better solution. In recent times, the decisions reached have to an increasing degree favoured the importation of high-grade ore.

Where inexpensive water power is available—as in Norway, Rhodesia, Venezuela and so on—the utilisation of electric smelting processes, or the establishment of other industries using electric power (aluminium works, for example), is an obvious step to take.

Natural gas, too, is a very important raw material today. It secured its position in the USA power industry a long time ago, but has only recently entered into competition with orthodox media in Europe. Favourably-priced electric power can in many cases be generated using modern natural-gas turbines, and the possibility of using natural gas as the sole reducing agent in the smelting of iron ore will be discussed later in this book. Not only have high-capacity metallurgical plants been built in countries which have inexpensive natural gas at their disposal (Mexico, for example), but numerous projects are the subject of intensive planning at the present time in other countries that have their own rich natural gas deposits, in the Near East, for instance, the object being the establishment of iron and steel industries on the specific basis of natural gas. Of course, natural gas can also be used in other branches of industry. In the iron and steel plant itself, natural gas can also be used in a secondary role as an inexpensive source of heat for the numerous auxiliary heating appliances, e.g. for ladle-heating facilities, rolling mill furnaces, and many other heating operations.

The presence of oil is also of interest. Not only can it be used for the inexpensive firing of furnaces, and possibly for power generation, but thoughts are now turning to oil injection in classical blast-furnace processes too, with the object of saving a certain proportion of expensive coking coal or coke. Finally, attention is drawn to the possibility of using oil refinery by-products such as petroleum coke as a basis for electric metallurgical processes.

Other raw materials and their locations play but a minor role in comparison to ore and coal. They are normally to be found everywhere without undue difficulty, and include limestone, dolomite, additions, building materials and so on. Consequently, these materials are left until the actual planning work is undertaken.

III. Plant Location and Transportation Facilities

As intimated in the preceding section, the location of the raw material deposits is not the sole governing factor for plant site selection; the transportation problem exerts considerable influence, too. Given good water transport conditions, foreign ores delivered free works can be cheaper than domestic ores that have to be transported by rail, an expensive means of moving ore.

a) Rail Transport

In spite of the fact that water transport is cheaper and particularly suitable for bulk goods, the railway is still the most important means of transportation in this field. Waterway freeze-ups lasting several weeks must always be expected in winter—at least in Europe and North America—when the railway must bridge the gap. The efficiency of the railway must be judged in this light.

Consequently, metallurgical plant must be so linked to the general railway system that the delivery and despatch of *all* materials needed and produced by the works can be dependably effected. Rail transport facilities must exist for the volume of goods normally handled by waterway or road transport, too, in case the latter facilities are brought to a standstill by weather conditions (as mentioned above). Suitable branch lines and loading facilities must also be provided.

The situation caused by the present division of Germany provides us with a striking example of how problematic the transport situation can become. The West German iron and steelworks situated near the Iron Curtain have lost their most favourable market in Central Germany. In consequence, they have also lost their not inconsiderable advantage over the freight bases of Dortmund, Oberhausen and so on as regards freight conditions, and must at present market part of their output through the appropriate freight bases on less favourable terms.

During all tariff negotiations, special attention must be paid to the specific railway conditions in the country in question as regards demurrage. If the railway authorities have unrealistic demurrage conditions, i.e. if the time the railway wagons are allowed to remain in an iron and steelworks is too short, the sums to be paid in demurrage charges may well prove to be insupportable. The conditions for feeding in materials by railway wagons and for the removal of wagons laden with finished products must therefore be given careful consideration.

The frequent practice of utilising the empty wagons that brought in bulk goods for the onward transport of finished goods is, of course, in itself a practical arrangement and in the best interest of the national economy. However, the period the wagons may be retained before charges are raised must be such that the plant can operate normally without running the risk of having to pay demurrage.

Difficulties connected with the discharging of wagons under adverse climatic conditions, e.g. unloading frozen ores in winter, must be given due consideration from the very outset from the demurrage aspect. This applies in particular to countries in which during a period of many months the temperature varies greatly above and below 0 °C at rapid intervals. In such cases, in order to regulate the demurrage situation and also the actual operation of the plant, suitable facilities such as thawing bays must be provided for, unless these difficulties can be obviated by other means.

b) Inland Waterways

As water transport is particularly suitable for bulk goods such as coal, ore, limestone, building materials, etc., and for rolled products, the siting of a new iron and steelworks near a waterway is most desirable whenever possible. At the same time, every effort must be made to establish a works-owned harbour in order to permit unloading and loading to be effected without transshipment being necessary, the goods delivered free bunker or supplied ex-works being handled direct. It must always be borne in mind that each and every transshipment of bulk goods, such as ore, for example, involves not only handling costs, but a loss of substance, too. This often accounts for 2% *and more* of the amount handled per transshipment operation.

The costs involved when building any type of hydraulic structures, and especially when locks and so on have to be built, are often disturbingly high. However, as an iron and steelworks always provides a uniform and significant volume of goods to be handled on the waterway, ways and means of obtaining relief in the financing of such projects can be found in most cases. As most waterways are state-owned, it will in many instances prove possible to have all or part of the costs of such connecting installations taken over by the state.

c) Sea Transport

Sea transport can only be utilised by new iron and steelworks located in seaports or connected to inland waterways. Here, too, the possible need to transship from ocean-going vessels to barges must be taken into consideration; works situated on waterways that can be navigated by ocean-going vessels—the Lower Rhine and the Thames—enjoy the advantage of direct supply without transshipment to barges.

d) Motorways

European conditions scarcely permit the transport by road of
the bulk goods needed by an iron and steelworks, or of the work's
products, but in many other countries the situation is different.

For example, in many cases calculations prove that in countries
such as Persia and America, which produce their own petrol, road
transport by lorry is cheaper than rail or any other form of transport,
even for great volumes. Of course, good connections to the main
network of roads and motorways are indispensable for a new
industrial centre. Practical connections to neighbouring towns and
cities are also of great importance and can in certain cases save
considerable sums that would otherwise have to be invested in the
infrastructure sector when establishing domestic iron and steel works.

e) Air Traffic

If the nearest civil airport is too far away from the plant, early
provision must be made for an airport with good connections to the
road network. The various aspects of the prevailing wind (dust)
that affect residential areas etc. also affect the siting of the airport.
In addition, the plant may not be situated in the main aircraft
take-off lanes or landing approaches. The question of whether and
to what extent the airport may later be extended to cater for
regular public and postal services must also be carefully examined.
Even if only a slight probability of this happening exists, it must
be taken into due account.

IV. Labour Market Conditions

Normally, non-industrialised areas have no skilled industrial
workers. The execution of a plan of this nature therefore means
that one has to accept the fact that a certain proportion of skilled
tradesmen *must* be brought from other areas. It must then be
established whether the large number of unskilled workers required
is available in the district, whether or not their employment is a
practical proposition, and whether or not the character of the population makes this possible. Further investigations must reveal
whether this is feasible at least to a limited extent, and, finally,
how many unskilled workers must be brought in from other areas.

The analysis of the labour market must then be extended to
cover such questions as to where these workers can and should be

obtained from. Experience gathered during the past few decades —and particularly in Germany in 1937/39, quite apart from the war years — has more than proved the paramount importance of an analysis of this nature. Care and thoroughness in this respect cannot be overdone.

An equally important question is the provision of suitable temporary accommodation for this labour force, followed by the establishment of permanent housing estates with all usual amenities.

When discussing in this section the obtaining of skilled tradesmen from other areas, we are naturally talking in terms of temporary employment only. One cannot say of any country in the world that the inhabitants are not intelligent enough to operate even modern industrial plant. The experience of the authors shows that even in countries which hitherto had no industrial potential whatsoever, it proved possible to leave the running of the new industrial plant to the inhabitants after a more or less short period of time. Of course, this calls for the thorough training of unskilled local workers in the countries supplying the plant and equipment in due time before the commissioning of the new iron and steelworks. In some instances, it proved possible to completely hand over an iron and steelworks to the new personnel within twelve months in countries that to then had no iron and steel plant at all; this naturally called for sound and thorough training. In general, it may be said that provided suitable preliminary training is given, it should prove possible to enable the local crew to operate the plant on their own, or at least with only a very limited number of specialists from other countries, within a period of perhaps three years.

V. Air Pollution Control

Each and every industrial concern that discharges gases or dust of various types into the atmosphere is obliged by legislation in force in all industrialised countries to adopt certain measures to prevent the discharge of these gases and dusts, or at least reduce the volumes discharged to the lowest possible level.

However, the waste gases and dust that still enter the atmosphere in spite of the measures adopted can have undesirable and even damaging effects in the vicinity of the works where larger volumes are concerned. The larger the plant, the larger the volumes of gas and dust involved per hour or day.

For these reasons, the iron and steelworks must from the very outset purchase a large proportion of the surrounding area to eliminate the risk of claims for damages being made at a later date.

In the following, we summarise the results of investigations undertaken to ascertain what minimum area of ground must be purchased by the works to prevent claims being raised for damage caused by gas and dust.

The figures quoted originate from planning work and calculations made during the building of a large iron and steelworks, as well as a continuous series of measurements taken over a period of 13 months while the plant was in operation to establish the volume of dust falling on the works and the immediate vicinity.

Considerable volumes of gas and dust are discharged into the air by the iron and steelworks, its power station, and its ore preparation plant. Table 1 summarises the hourly rates of waste gas and dust, together with the compositions of the dusts emitted by the various main departments of the works. All figures quoted are based on an annual production rate of approximately 1 million tons of crude steel. These dust calculations presuppose the correct functioning of the prescribed dust-extracting equipment used, which includes both mechanical and electric facilities.

In most cases, the dusts are composed of *limestone, silica,* and *alumina,* together with *phosphoric acid* from the basic Bessemer plant; all these components are harmless in residential and agricultural areas. Only where the deposited volumes are really high does the dust constitute a nuisance and possibly a danger. The same applies to the soot discharged from the power station smoke stacks, the only difference being that soot is obviously much more of a nuisance in residential areas.

Alkalies, discharged in the main together with waste gases from the ore preparation plant, are on the other hand much more unpleasant; as they are soluble, they penetrate the ground when dissolved in dew or rain and can thus affect plant growth.

The same applies to sulphurous acid. The sulphur in the coal is released as SO_2 through the power station smoke stack; mixed with dew or rain, it produces sulphurous acid. The normal SO_2 content in power station waste gases lies below 0.1 g per Nm^3, and this low concentration is generally considered to be harmless. However, the use of coal with higher sulphur contents when the power station is working all out can produce a temporary multiplication of the stated SO_2 content.

Table 1. *Example of waste gas and dust rates*
(related to 1 million tons of crude steel per annum)

Plant section	Waste gas		Dust		Type of dust
	Nm³/h	%	kg/h	%	
Coking plant	230,000	6.3	5	0.5	Dense tarry smoke
Blast furnaces	400,000	11.0	20	2.0	Lime, silica, alumina, alkalies
Ore preparation plant	770,000	21.1	400	40.0	Lime, silica, alumina, alkalies, arsenic hydride, chlorides
Lime kilns	13,000	0.4	34	3.4	Lime
Steelworks	350,000	9.6	525	52.5	Lime, phosphoric acid, silicon
Rolling mills	130,000	3.6	—	—	(see "Blast furnaces")
Power station	1,750,000	48.0	16	1.6	Soot, SO_2, silica, lime
Totals:	3,643,000	100.0	1,000	100.0	

Content of the *full volume* of blast furnace gases, i.e. of the full volume burnt in all departments from the coking plant to the rolling mill.

The question of *arsenic* merits particularly close attention. Salzgitter ores, for example, have a minor content of arsenic, lying between 0.03 and 0.06%. In this specific case, this arsenic remains 100% in the ore on the sintering belt and is fed into the blast furnace, but the arsenic content in blown hot metal is so small that it has no adverse effects on the iron. On the other hand, if the sinter must be quenched on the sintering belt, arsenic can combine with hydrogen and be emitted in a gaseous form together with the waste gases as arsenic hydride. In this case too, it can be reduced to a watery solution by rain and so penetrate into the ground.

Finally, the formation of *chlorides* and their evacuation in waste gases must be expected; this, too, can cause damage to plant life. The use of high smoke stacks in the power station and the ore preparation plant (90—130 m) ensures from the very outset that the dust-laden waste gases are discharged into the atmosphere at a sufficient height from the ground to prevent excessive concentrations being deposited in the immediate vicinity.

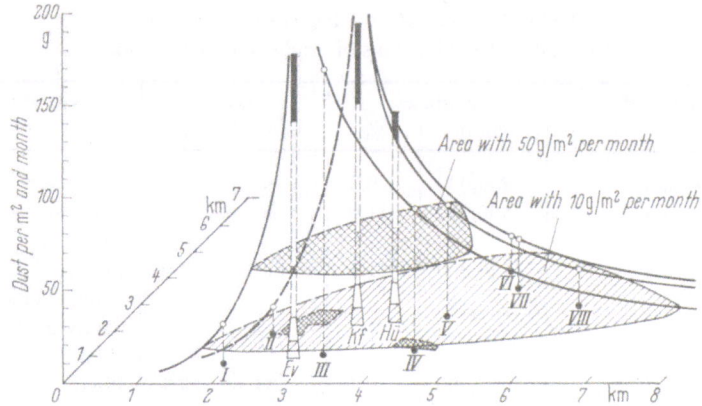

Fig. 1. Evaluation of dust volumes over a period of 13 months in the works area and near vicinity. Prevailing wind from WSW.

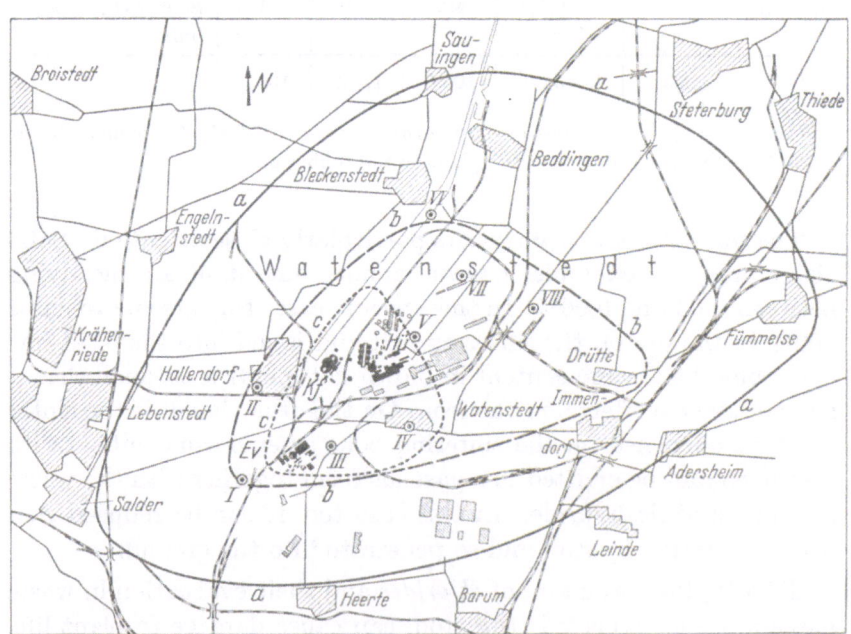

Fig. 2. Projection of the measured values.

a Extreme dust deposit limit; b limit of area with 10 g/m² per month; c limit of area with 50 g/m² per month.

In order to determine the actual effects of the dust deposits and
the position of the affected area, careful and accurate measurements were taken over a period of 13 months in eight different
locations in the vicinity of the iron and steel works. The only way
to obtain from these measurements a clear picture of the dust-
affected areas within the plant and in the surrounding area was to
draw up a three-dimensional chart as shown in Fig. 1. In order
to facilitate the reading of this chart, the areas with 50 g and 10 g
of dust per square metre per month are shown as two separate
sectional planes. These two projected areas are marked on the map
shown below (Fig. 2), together with the boundary within which
the volume of dust deposited can still just be measured. This latter
area was determined by repeated measurements taken after fresh
falls of snow, which permitted the exact determination of the
boundary.

This evaluation, taken together with the dust volumes listed in
Table 1, permits a more exact compilation:

Designation	Area in km²	Dust volume per month kg	g/m²
a) Total area	50.0	600,000	12
b) Areas with up to 10 g/m² per month	13.4	490,000	37
c) Areas with up to 50 g/m² per month	4.1	246,000	60
d) Area b—c	9.3	244,000	26
e) Area a—b	36.6	110,000	3

which reveals that the average dust volume in the overall area is
12 g per square metre per month, whereas for the innermost area
(c) the figure is 60 g, and in the intermediate area (d and e) 26 and
3 g. The shape and location of the overall area affected clearly con-
firms the direction of the prevailing wind, which in this district
has been WSW for many years.

Furthermore, the overall area of 50 km² is equivalent to a circular area with a radius of 4 kilometres, a value that has often
been established for similar industrial plant that discharges dust
into the atmosphere.

We may therefore confidently assume *that at least the overall area
of 50 km² shown in the map can be affected in one way or other by*

*waste gases and dust expelled into the atmosphere by the iron and steel-
works*, and that within this area damage can be caused by occasional
larger volumes of sulphurous acids, arsenic hydride, alkalies, or
chlorides, which are always possible; the damage so occasioned
affects agriculture in the main, and can lead to claims for damages
being made by the injured parties.

We have here discussed the types, volumes, and the significance
of waste gases, smoke, and dust discharged into the air by iron
and steel works; naturally, similar statements could be made in
regard to chemical works, power, stations and so on. Now the
degree of air pollution caused by a single works may not be excessive
but the combined effects of several works in industrial areas are
quite a different matter altogether. The canopy of dust and smoke
etc. over the Ruhr district, Germany's coal-mining and steel centre
(which is also the home of various chemical works, power stations,
and the multitude of other works that are established in such
industrial centres) has now become a menace to the health of the
population. The countless chimneys of London pour out smoke that
combines with fog to produce the notorious "smog" (smoke-fog)
which constitutes a similar threat to the health of all who live and
work in this city.

During the past few years, various laws on air pollution control
have been passed in the Federal Republic of Germany. They
demand that all waste gases etc. expelled into the air by power
stations, chemical works, coking plants, iron and steelworks and so
on be cleaned to the highest possible degree before being discharged.
Dedusting processes have now been developed to such a point that
the values given in the example quoted above and in Figs. 1 and
2 may well be reduced in the future.

VI. Agricultural Land

It is obvious that the agricultural industry views the use of arable
land for an iron and steelworks site with disfavour. In Europe,
where every square metre of arable land is precious, the release of
good land for industrial use will always have to be the subject of
very close scrutiny, and the final decision is then a matter for the
higher governmental planning authorities.

The acquisition of neighbouring land by the iron and steelworks
is a different matter altogether. Land subject to mining subsidence,

i.e. land above ore and coal workings etc., simply must be acquired; the costs occasioned by ground subsidence damage can reach incredible heights, and by purchasing the land the plant can also obviate an endless chain of vexatious negotiations. The immediate purchase of the land in the near vicinity is also an advisable step to take; one must also take into account the fact that the iron and steelworks will directly affect the surrounding area in one way or another — we are thinking here of dust, smells, noise, etc. — and by purchasing this land the works can avoid unpleasant complications with nagging neighbours.

Legislation exists covering damage caused by the mining industry, and most countries also have laws covering damage caused by smoke, dust, smells, etc., emitted from industrial works; consequently, disputes can become extremely unpleasant. This applies in particular to new industries in traditionally agricultural areas; here, the effects of any smoke, etc., harmless as they may be, are often (and quite understandably) viewed and treated as instances of "damage", whereas exactly the same things never give rise to complaints of any sort in traditionally industrial areas. Naturally, land required for later expansions of the works and for housing schemes, etc. must always be given due consideration during the initial planning stage.

The most profitable utilisation of the land is, of course, a prime consideration. Whether this is effected by the works itself through an estate administration organisation, or whether the land is leased in sections is immaterial; the main thing is that the land is put to the best possible use.

VII. Political Borders

Unfortunately, past experience has shown that as far as planning work and installations of this nature are concerned, provincial and county borders can produce obstacles that are not always easy to negotiate. Recognition of these difficulties and their fundamental clarification at the earliest possible stage of the planning work is advisable.

If the area occupied by the works and its housing estate assumes extensive proportions, any restrictions imposed by county, district, provincial or other boundaries are more than irritating. Essential dealings with authorities such as public building authorities, industrial authorities, land registries, political authorities, etc. are immediately doubled when the iron and steelworks lies in two counties, or, worse still, in two provinces. In these cases, every effort must be made to obviate the difficulties by shifting the boundaries.

Where national borders are concerned, the overall project must be coordinated by the governments concerned in any case.

VIII. Profitability Estimation

All industrial planning projects are based on estimated profitability calculations, or at least on rough estimates. The most important motives for the planning and the site selection are known, and it must also be possible to express these basic factors as costs. This applies to the costs per ton of ore, coal costs free works, i.e. coal costs including freight, and to the presumable sales returns for rolled products, slags, long-distance gas, electric power and so on.

Naturally, the exact production costs per ton of pig iron, crude steel, and rolled steel cannot possibly be dependably calculated in the preliminary planning stage, but an experienced industrial economist can draw up a rough estimate of the work's profitability status using the factors outlined above. Of course, care must be taken to exclude dubious or unfounded values from this calculation; however, if sound empirical values are taken as a basis, thorough and careful calculation can produce a useful and dependable result.

This view has been doubted in the recent past — at least in Germany — but the authors and other experts of repute are convinced of its value, which has also been proved during the establishment of new iron and steelworks in Europe in recent years.

Profitability calculations are covered in detail on page 171.

B. Technical Fundamentals

This section describes the technical fundamentals and operating procedures that form the basis of all planning work in respect of iron and steelworks. In addition to the fuels available in the area in question, the nature of the available iron ore is the main governing factor when selecting processes for the reduction of ore to pig iron, for steelmaking, and to a certain extent for the further processing of the steel to the usual end products of a normal iron and steelworks.

I. Fuels

The past few decades have witnessed a pronounced change in the iron and steelworks fuel sector. With the exception of coke, which is still indispensable in blast-furnace operations, the earlier extensive use of coal for the generation of producer gas and for firing half-gas furnaces and steam boilers has sharply decreased; in turn, the consumption of rich gases — coke-oven gas, natural gas, etc. — and heavy fuel oil has increased to a marked degree.

a) Solid Fuels

In spite of vigorous research and development work aimed at the production of iron outside the blast furnace by using various processes as described later in this book, the time-honoured coke blast furnace is still the mainstay of iron ore smelting operations, and will retain this status in the foreseeable future. Consequently, *coke* will continue to be used as a smelting fuel during the coming years, probably for decades.

The bulk of coke-oven gas is produced in orthodox ovens. Coking plant design is mainly determined by the type of coal available; not only do the coals mined in different coalfields (Ruhr, Saarland, Upper Silesia, Belgium, Wales, etc.) differ greatly in nature, making coke ovens of various designs necessary, but even the kinds of coal mined in the one coalfield often vary to a marked degree. If it is known from the beginning that only one or a limited number of exactly specified kinds of coal is to be used in the coking plant,

2*

planning work is greatly simplified. However, because of present-day large-scale planning schemes and the control of the coal industry by bodies such as the High Authority of the European Coal and Steel Community, one cannot rely on the bulk or entire quantity of coking coal required being supplied from the same collieries — as far as the country in question has its own coal-mining industry. It is therefore recommended that provision be made for the smooth processing of different and also *varying* kinds of coal; where the existing communications mean that an iron and steelworks can be supplied with coking coal from any one of several coalfields, this precaution is even more of a necessity. The main requirement is for a coal blending and crushing plant which can cater for the largest possible number of different kinds of coal; the appertaining stock-yards and bunkers must be sub-divided as required.

The coke ovens should be matched in size and design to the kinds of coal available; if, for example, a coal rich in gas is to be coked (e.g. Saarland and Silesian coals) the ovens must be designed accordingly, i.e. the retorts, or chambers, will have to be wider than usual, 500 mm generally being the lower limit. As coals of this type cannot be heaped on, but have to be tamped down, the chamber height may not exceed a certain value. Coke ovens fed with Ruhr coal have been designed with chambers 6 metres in height, but 4 metres is the normal figure. The length of the chamber is not so important and generally lies between 10 and 30 metres.

As far as oven arrangement is concerned, the maximum number of ovens to be served by one coal storage tower is about 200 to 220. Bigger batteries cannot be served by a *single* coal storage tower. The siting of coal storage tower and coking ovens in relation to one another is otherwise mainly governed by the local planning arrangements and the location of the coal feed point, i.e. the coal blending and crushing plant. The coke pushers and coal charging cars also serve a limited number of ovens, which means that the most practical solution for large batteries with up to 200 ovens would be to site the coal storage tower in the centre of the battery, and to use one charging car and one pusher on each side.

Iron and steelworks coking plants are invariably equipped with combination ovens; the system selected from case to case — i.e. either Didier, Koppers, Otto, Still, etc. — depends on the existing business connections and the taste of the persons in charge of planning. All German oven types have given excellent service; they differ in design and mode of operation, but are fully equal as regards their practical operational results and dependability, which makes it impossible to say that the one or other system is superior to the rest. All coke oven operators have gathered wide experience in the coking of all German and most foreign kinds of coal, but it still cannot be said that any one system is particularly suitable for coal from specific coalfields.

Coke quenching plants are nearly always of the wet type. Even though dry quenching has made advances in recent times, its practical application is still accompanied by considerable difficulties. The addition of a separate boiler installation represents a not inconsiderable complication of the entire plant; again, although dry quenching produces an almost absolutely dry coke, one has to accept an additional coke loss by combustion in the order of 2%. In addition, the coke presents an ugly surface with an impaired reaction characteristic. All in all, in the majority of cases orthodox wet quenching will therefore be preferred to dry quenching.

The *by-product recovery plant* will be designed on tried and tested principles; the only decision to be taken is whether liquid ammonia is to be produced in the usual form, or as a concentrate, or ammonium sulphate, in addition to the normal recovery of tar, benzene, and sulphur. In coking plants that supply long-distance gas, the benzene recovery plant and the desulphurisation plant are nowadays of high-pressure design, i.e. the long-distance gas has to be compressed in the heart of the by-products recovery plant. The high-pressure benzene recovery and high-pressure desulphurisation system offers certain advantages, chief among them being on the building side, where a substantial saving is made in structural steelwork and space.

The *compressor plant* required for long-distance gas supplies, the desulphurisation plant, naphthalene extraction plant, and gas-drying plant are of the usual design. Because of the present-day long-distance gas pressures of up to 50 atmospheres, taken together with the flexibility now demanded of the gas supplier, the discussion as to whether turbo compressors or reciprocating compressors are more suitable for long-distance gas installations has in the majority of cases been decided in favour of the multistage reciprocating compressor. Of course, both turbo compressors and reciprocating compressors can be installed, the former for the base load and the latter for the peak loads, but the compressor man will not always be keen on operating a mixed plant of this sort, quite apart from the adverse effects the piston movements of the reciprocating compressors can have on the turbo compressors. However, the problem of pulsating gas columns in the pipework can nowadays be adequately countered.

As opposed to the practice of earlier years, the lump size of blast furnace coke has in recent times been matched to the uniformly prepared burden, i.e. instead of the lump size of 80 mm (3″) and more usual in earlier days, coke with lump sizes of down to 40 mm (1½″) is being charged into blast furnaces on an increasing scale. The small coke and coke breeze accumulating during crushing are largely used on the sintering strands, which are increasing in both significance and number. The normal quantity of small coke available is nowadays often insufficient to meet the requirements of even the sintering plants.

Pit coal is virtually the only other solid fuel used in the classical iron and steelmaking countries, whereas brown coal — lignite — is still often used in other countries. The earlier widespread practice of using coal in grate-type or half-gas fired rolling mill furnaces has practically disappeared from the iron and steelmaking scene; in isolated instances, furnaces of this type are still fired with coal dust, which is either supplied in special wagons or ground in blowing-type mills and injected by these into the furnaces. The use of producer gas has also decreased to a marked extent; gas producers are nowadays used only for the production of clean fuel gas from small coke for the undergrate firing of coke ovens. The only remaining coal consumers in iron and steelworks are the steam boilers; normally the coal is here fired as coal dust, and to a lesser extent on travelling grates together with blast furnace gas.

b) Gases

1. Blast Furnace Gas

Blast furnace gas is still the backbone of the fuel supply system of every integrated iron and steelworks. Some 4,000 Nm³ of blast furnace gas can be expected per ton of blast furnace coke used; about 6—10% of this gas volume is unavoidably lost, leaving 3,800 to 3,600 Nm³ of *useful* blast furnace gas at the work's disposal.

Now that a series of measures such as ore burden classification, the use of richer ores, higher proportions of sinter in the burden and so on has greatly reduced the coke rate in kg/t pig iron, while at

Table 2. *Examples of blast furnace gas balances*

Example	1955				1964			
	Ia		Ib		IIa		IIb	
	Nm³/t^a	%	Nm³/t^a	%	Nm³/t^a	%	Nm³/t^a	%
Coke oven plant	600	24.7	—	—	285	19.5	—	—
Blast furnace (incl. sintering plant etc.)	770	31.8	770	31.8	565	38.8	500	34.2
Gas turbo blower	—	—	—	—	135	9.2	135	9.2
Steelworks	255	10.5	255	10.5	30	2.0	100	6.9
Rolling mills	400	16.5	400	16.5	330	22.6	360	24.7
Other consumers	55	2.3	55	2.3	20	1.4	50	3.4
Power plant	345	14.2	945	38.9	95	6.5	315	21.6
Total:	2,425	100	2,425	100	1,460	100	1,460	100

Explanation		Examples Ia and Ib	Examples IIa and IIb
Gas calorific value	kcal/Nm³	950	875
Coke ratio per ton pig iron	kg/t p.i.	900	635
Gas yield per ton of coke	Nm³/t coke	4,000	4,000
Available gas volume per ton			
of coke	Nm³/t coke	3,740	3,740
of pig iron	Nm³/t p.i.	3,366	2,375
Gas-kcal per ton of pig iron	10³ kcal/t p.i.	3,198	2,078
Pig iron per ton of crude steel	kg/t steel	759	703
Gas-kcal per ton of steel	10³ kcal/t steel	2,425	1,460

^a Converted to Nm³ with a calorific value of 1,000 kcal per ton of crude steel.

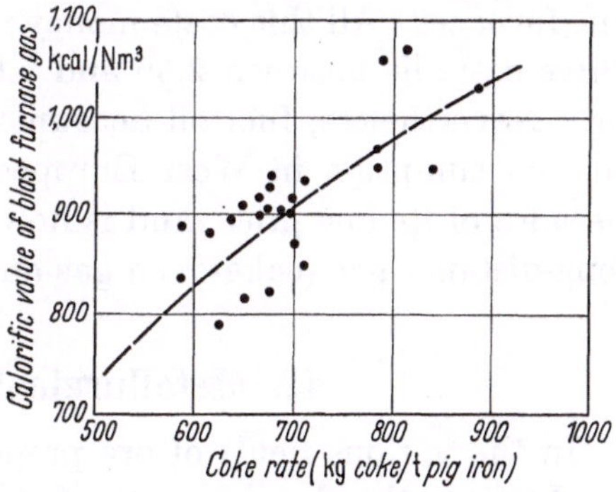

Fig. 3. Calorific value of blast-furnace gas (kcal/Nm³) as a function of the coke rate (kg coke/t pig iron).

the same time the utilisation of blast furnace gas has been improved by more intensive indirect reduction, the ratio of CO_2 to CO in the blast furnace gas has shifted in favour of a higher CO_2 content, bringing about a considerable drop in the calorific value of the gas as compared with earlier days, a drop that may perhaps become more pronounced as time goes on (Fig. 3). Table 2 shows two blast-furnace gas balances dating from 1955 and 1964 that show in some measure the tendency of the change described above.

2. Rich Gases

This term includes both coke-oven gas and natural gas. Long-established practice puts the gross calorific value of coke-oven gas at 4,300 kcal/Nm³, which equals a net calorific value of 3,800 kcal/Nm³. Depending on its place of origin, the gross calorific value of natural gas lies between 8,500 to almost 10,000 kcal/Nm³, giving a net calorific value of between 7,500 to almost 8,800 kcal/ Nm³. For rough calculations, the gross and net calorific values of natural gas may be taken as being twice as high as those given above for coke gas.

Coke-oven gas — and lately in Western Europe natural gas, too — occupy an important place in the fuel systems of iron and steelworks. They are easy to convey, are clean to use, and permit very exact setting of burners and furnaces.

c) Fuel Oil

Special cases apart, iron and steelworks use heavy fuel oil only. The volume of tar-oil won as a coking plant by-product is very small indeed, and the bulk of the fuel oil requirement must therefore be covered by mineral oil. This fuel oil is highly viscous and must be heated to about 60 °C for pumping, and to 120 °C for injection into the furnaces. All this costs money; in German iron and steelworks, these costs lie between 2.50 and 7.50 $ (10 and 30 DM) per ton of oil. Nevertheless, fuel oil accounts for a major proportion of the fuel consumption in West European iron and steelworks, mainly because of its low price, and is now offering serious competition to long-distance gas (coke-oven gas and natural gas).

II. Metallurgical Processes

In the various fields of ore preparation, ore smelting, and steel production, the development of new processes long in the offing has in recent years been greatly accelerated; these processes will therefore be given particular attention in this section.

a) Ore Preparation and Ore Beneficiation

The term "ore preparation" describes the physical preparation of the ore by crushing and screening, and agglomeration by sintering or pelletising, whereas the term "ore beneficiation" (or "dressing") describes treatment of the ore by chemical and mechanical processes to enrich the Fe content and remove undesirable components such as silica, alumina, limestone and so on. Ore beneficiation involves the use of various processes such as flotation and magnetic separation. However, as the two treatments often overlap, the main processes are described together in the following.

As early as thirty years ago, leading blast-furnace engineers such as v. REICHE, KINTZINGER, A. WAGNER, LENNINGS, SENFTER, the past-master of the art HERMANN RÖCHLING, and the American H. A. BRASSERT, emphasized the need for extensive ore preparation in German blast-furnace plants, and eventually had their way.

The realisation that even low-grade ores as available in great quantities in the Salzgitter, Dogger, Minette, and Corby (England) fields could and must be smelted on a large scale, first dawned in 1936, quickly gained ground, and led the German iron works to expand their ore preparation and beneficiation plant; this development was accelerated by wartime demands.

In the USA, too, depletion of the Mesabi ore deposits long ago led to the exploitation and dressing of taconite, which has a low Fe content.

The Studiengesellschaft für Doggererze (German Association for Dogger Ores) has developed various ore beneficiation processes which have proved particularly suitable for lean ores. The installations were developed in the "Dogger" district (Germany), and a major facility was erected at Salzgitter in 1940. Two distinct processes are described below; firstly the wet process, and secondly the dry process.

1. Wet Process

The installations developed for Salzgitter ores are based on the washing method, a certain content of alumina and silica being the prerequisite for the application of this process. The large-scale facilities at Salzgitter have an annual crude ore capacity of 6 to 7 million tons and have given every satisfaction; they produce a good yield at low production costs. Fig. 4 shows a quantitative flow sheet of the Calbecht wet dressing plant[1].

The wet concentrate has a moisture content of 12 to 15%. In most cases, this is not a disadvantage, as the wet concentrate nearly

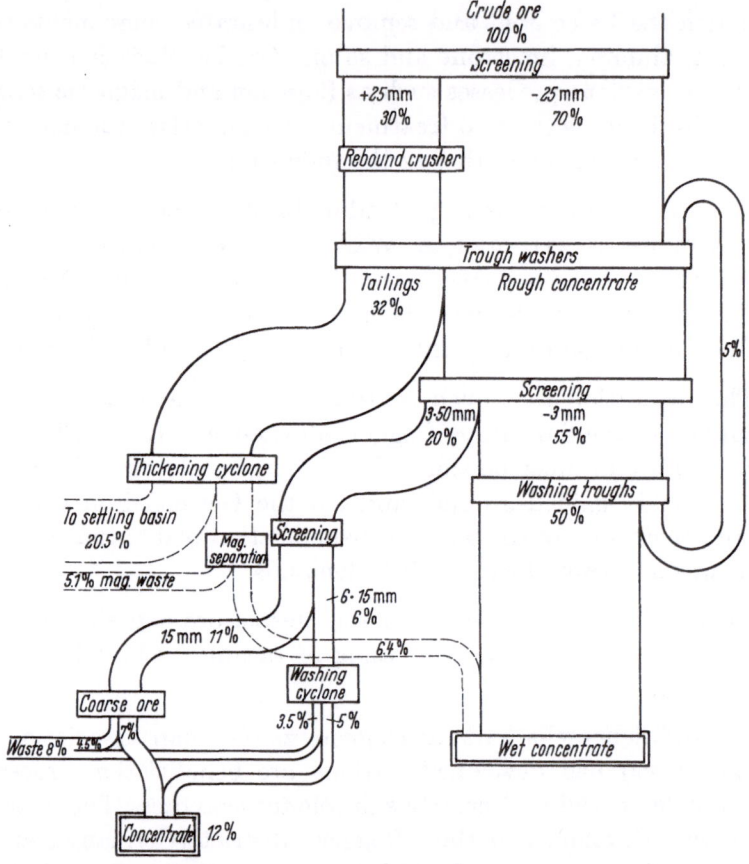

Fig. 4. Quantitative flow sheet of the "Calbecht" wet beneficiation plant (by A. COLTZ and K. NEUMANN).

always comes into contact with other small ore, blast furnace flue dust, and dry concentrates on the sintering strand, which means that the latter need not be wetted separately. As far as acid and phosphoric ores are concerned, a decided advantage of the wet

---

[1] GOLTZ, A., and K. NEUMANN: Die Erzaufbereitungsanlage Calbecht der Erzbergbau Salzgitter AG, pp. 249/59 (The Calbecht Ore Dressing Plant of the Erzbergbau Salzgitter AG).

mechanical process is that some 33% of the SiO_2 is removed and the main bulk of lime and phosphorus remains in the concentrate; at the same time, the crude ore is enriched by 10 to 15 points Fe, i.e. a crude ore with 27 to 37% Fe produces a concentrate with

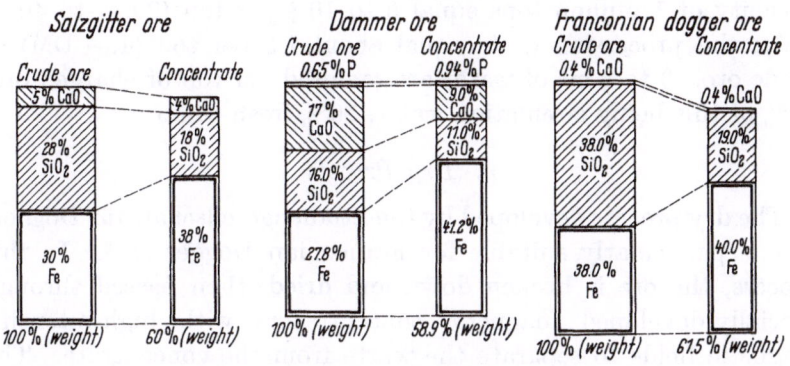

	Crude Ore	Concen- trate	Crude Ore	Concen- trate	Crude Ore	Concen- trate
Yield (quantity)	100%	60%	100%	58.9%	100%	61.5%
Yield (Fe)	100%	76%	100%	87.1%	100%	85.0%
SiO_2 expelled	—	63%	—	59.5%	—	69.2%
Fe/SiO_2	1.07	2.11	1.74	3.75	0.76	2.11
CaO/SiO_2	0.18	0.22	1.06	0.82	0.01	0.02

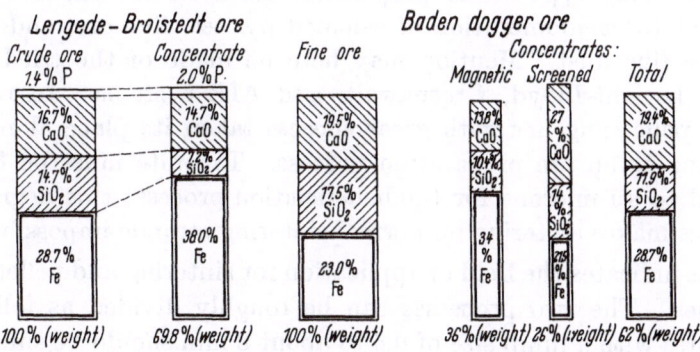

	Crude Ore	Concen- trate	Fine Ore	Mag. Concen- trate	Screened Concen- trate	Total Concen- trate
Yield (quantity)	100%	69.8%	100%	36%	26%	62%
Yield (Fe)	100%	92.4%	100%	53%	24.3%	77.8%
SiO_2 expelled	—	65.8%	—	57.7%	—	57.7%
Fe/SiO_2	1.95	5.29	1.31	3.27	1.54	2.41
CaO/SiO_2	1.17	2.04	1.11	1.83	1.93	1.63

Fig. 5. Characteristic values of German ore beneficiation plants (by Studiengesellschaft für Doggererze) [All analyses refer to dry matter].

nearly 40% Fe. Fig. 5[1] shows the results obtained with this beneficiation process. This chart deals with both wet and dry processes,
but only Franconian Dogger ore is taken as an example for the wet
process.

The investment costs for a large installation with an annual input
capacity of 1 million tons equal 5 to 10 $ per ton (20.— to 40.—
DM); the processing costs equal about 1 $ per ton (4.— DM) of
crude ore. 3 to 4 m³ of water are required per ton of charged ore,
80% of this being circulation water, 20% fresh water.

2. Dry Process

The dry process developed by the Studiengesellschaft für Doggererze is particularly suitable for Franconian Dogger ore[2]. In this
process, the ore is broken down and dried, then passed through
specially-developed magnetic concentrators with high-intensity
magnetic fields to separate the waste from the concentrate. The
investment costs are in the same order as those of wet plants, the
processing costs vary from 1 to 1.5 $ per ton (4.— to 6.— DM), fuel
costs forming the bulk. Power consumption equals some 10 kWh
per ton of crude ore.

Sintering and pelletising plants For a long time, sintering plants
were the only type of ore preparation facilities known, and were
used for the agglomeration of calcined pyrites, fine ore, and blast
furnace flue dust. Mention may here be made of the tried-and-
tested Dwight-Lloyd, Greenawalt, and AIB systems. In recent
years, pelletising has with great success taken its place alongside
sintering as an ore preparation process. Taconite must be finely
ground to 40 microns for the beneficiation process; this degree of
fineness makes sintering on normal sintering strands impossible.

This indicates the field of application for sintering and pelletising
facilities. The two processes can be roughly divided as follows:
Fine ores with a lump size of 0.5 to about 8 mm should be sintered,
a certain proportion of fines below 0.5 mm being admissible, of
course. All fine ores between 0 and 0.5 mm must be pelletised, and
the proportion of fines measuring more than 0.5 mm must be kept

---

[1] Provided by the Studiengesellschaft für Doggererze. See also ERICH BÖHNE:
Die Bedeutung der Aufbereitung einheimischer Eisenerze für die westdeutsche
Eisenindustrie (Importance of Dressing Indigenous Iron Ores for the West German
Iron Industry). Stahl und Eisen 77 (1957) pp. 549/52.

[2] See footnote b to Table 3.

to an absolute minimum; with certain kinds of ores, fines above
0.5 mm are not admissible at all.

Table 3. *Characteristic data of various ore beneficiation processes and sintering*[a]

Item	Unit	Studiengesellschaft Processes		Roasting[d]	Sintering[e]
		Dry[b]	Wet[c]		
Yield					
Concentrated roasted ore/ sinter per ton of crude ore	kg/t	640	630	588—775	750—900
Fe yield	%	84	81	94.3—95.6	—
Metal yield (Fe + Mn)	%	—	—	94.5—95.8	—
SiO_2 removal	%	75	57	65.8—81.0	—
Expenditure of:					per ton of sinter
Working hours	h/t	0.5	16.5	11—16	—
Electric power	kWh/t	10	4.5	8.5—11	45—50
Coke	kg/t	—	—	—	80—120
Water	m³/t	—	3	—	0.5
including: of fresh water (all per ton of charged material)	%	—	20—30	—	—
Operating Costs related to charge	$/t	0.75—1.0	0.40—0.60	1.10—2.0	—
	DM/t	3—4	1.50—2.50	4.50—8.0	—
related to sinter	$/t	—	—	—	3.75—5.0
	DM/t	—	—	—	15.0—20.0

[a] Data taken from "Erzaufbereitungsanlagen in Westdeutschland" (Ore Beneficiation Plant in West Germany), published by the Fachausschuß für Erzaufberei-
tung der Gesellschaft Deutscher Metallhütten- und Bergleute e. V., Clausthal-
Zellerfeld, Berlin/Göttingen/Heidelberg: Springer 1955.

[b] XIV. SENGFELDER, G., and A. GOLTZ: Die Aufbereitung der sandigen Doggererze in Nordbayern. (The Dressing of Sandy Dogger Ores in Northern Ba-
varia). — Die Aufbereitungsanlage für sandige Doggererze der Gewerkschaft
Eisensteinzeche "Kleiner Johannes", Pegnitz (Ofr.), pp. 225/32 (Plant for Dressing
Sandy Dogger Ores — Gewerkschaft Eisensteinzeche „Kleiner Johannes", Pegnitz, Upper Franconia).

[c] XVI. GOLTZ, A., and K. NEUMANN: Die Erzaufbereitungsanlage Calbecht der
Erzbergbau Salzgitter AG., pp. 249/59 (The Calbecht Ore Dressing Plant of the
Erzbergbau Salzgitter AG).

   ᵈ XVIII. GLEICHMANN, H.: Die Aufbereitung der Siegerländer Erze am Beispiel der Eisenerzgruben Füsseberg-Friedrich Wilhelm, Georg und Neue Haardt, pp. 266/313 (Dressing of Siegerland Ores — the Füsseberg-Wilhelm, Georg, and Neue Haardt Iron Ore Mines).

   ᵉ XVII. GOLTZ, A., and K. NEUMANN: Die Erzvorbereitung Watenstedt der Erzbergbau Salzgitter AG, pp. 260/65 (The Watenstedt Ore Preparation Plant of the Erzbergbau Salzgitter AG).

3. Sintering Plant

The various types of sintering strands and pan facilities on the market—Lurgi (Dwight-Lloyd), Greenawalt and AIB — are of differing designs but nevertheless work on virtually the same principle. In both cases, the fine ore is first mixed with 6 to 10% of coke breeze and spread upon the grate of the pan or the sintering strand, where it is ignited with blast furnace gas; air is sucked downwards through the ore-coke bed and sintering takes place. The ore particles are fused, and the resulting agglomerate is known as "sinter". Acid ores demand a certain proportion of limestone in the burden; KINTZINGER, BRASSERT et al established the fact that the addition of raw limestone to the fine ore prior to its being spread on the sintering strand is beneficial, and this practice has given good results. A further advantage lies in the fact that the coke requirement for melting limestone is substantially lower when the limestone is combined with the ore prior to being charged into the blast furnace.

Lurgi sintering strands are the most popular for high ore throughputs, the lower limit for economic operation being 1,000 t/24 h. The AIB plant, in which the pans are placed at different locations for igniting, sintering, cooling, and then tipping, whereas the old Greenawalt plant features 4 pans in turntable arrangement, offers the advantage that the sintering operation can be better regulated to suit the particular type of ore being treated; in addition, the pans can be replaced individually when worn or defective without the entire plant coming to a standstill. Dust development is a very unpleasant feature of all sintering plants, and this problem has not yet been fully solved, even with the most modern facilities.

Modern sintering plants can handle a throughput of up to 30 tons per square metre of treatment surface in 24 hours. A complete sintering plant designed for a throughput of 1 million tons p.a. (charge) can be estimated to cost 7.50 — 10 (30 to 40 DM) for every ton handled annually (these figures are based on conditions obtaining in 1966).

4. Pelletising Plant

The pelletising process has been used in the USA and Sweden since about 1935. Simply stated, in this process fine ores of 200 mesh (below 0.076 mm) are first mixed, then wetted, and finally rolled into pellets measuring some 10—20 mm (4—8″) in diameter on an inclined rotary disk or in a revolving kiln. Especially in the case of high-grade ores with a low gangue content, the firmness of the pellets is often increased by adding a binding agent, usually bentonite (about 2%). The green pellets are then calcined in a shaft furnace, on a sintering strand, or in a combination of sintering strand and rotary kiln (grate-kiln) at a temperature of 800 to 1,000 °C. The exact temperature is adapted to suit the ore composition. The nature of the shaft furnace restricts the capacity of a single unit to a maximum daily throughput of 1,000 tons, whereas belt or grate-kiln pelletising plants can be built for throughputs of up to 5,000 tons per day. Furthermore, the shaft furnace is unsuitable for baking pellets made of haematitic ores, the danger of caking in the sintering zone being too great.

In contrast to sintering plants, virtually all pelletising plants are nowadays erected and operated at the ore mines.

5. Ore Beds

In Germany, it was BRASSERT who first suggested using the ore bedding process to obtain a uniformly mixed burden, and he established the first ore beds in Salzgitter. These ore beds measure some 20×50 m, and each can take 3,000 tons of ore. A movable conveyor belt running above the ore bed feeds a continuous stream of sinter or ore along the entire length of the bed; the one basic essential is that the required volumes of the various ores are uniformly fed one after the other along the entire length of the bed until the latter is full. Distribution over the bed cross-section is immaterial. Uniform mixing is attained when the ores are reclaimed by the ROBINS rake, which roughly resembles a harrow; this rake slices cross-sections off the heap, the ore so removed falling onto an underground conveyor belt. The ore retains its uniformity of composition from the commencement of reclaiming to the end. If two or three grain sizes are used in the ore preparation plant, then two or three ore beds plus the necessary reserves will be sufficient. An ore bed installation for a throughput of 500,000 t of crude ore p.a. costs about 2 $/t (8 DM).

Fig. 6 shows the flow sheet of an ore preparation and beneficiation plant.

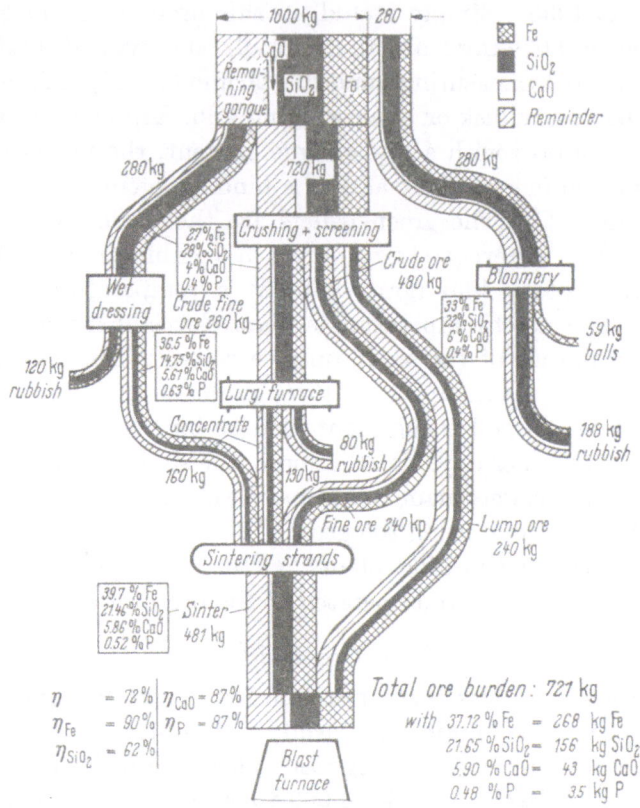

Fig. 6. Material flow — Salzgitter ore beneficiation plant (1940) crude ore with 29.88% Fe, 25,12% SiO₂, 4,96% CaO, 0,40% P.

b) Ore Reduction

1. Direct Reduction

The term "direct reduction" is used to describe processes to which a product similar to steel (as regards composition analysis) is obtained without first making pig iron to either a blast-furnace or an electric process.

In view of the present state of development of the direct reduction processes, this product is not available as ingots or billets and consequently cannot be put through the rolling mill direct; in the main, it takes the form of granules, pellets or briquettes and is therefore normally used as a replacement for steel scrap for charging open-hearth and electric arc furnaces. The fact that these products, also called "sponge iron products", are free from impurities such as chromium, nickel, arsenic, etc., (this naturally applies only when suitable ores are used), leads to the suggestion that sponge irons of this type may also be described as "high-grade scrap", as in many cases it is an ideal raw material for the production of high-grade and special steels.

Direct reduction processes use neither coke nor coking coal as fuels or reduction media; the following description of the processes is split into two parts to coincide with the fuels and reduction media normally used, i.e. on the one hand the gas reduction processes, and on the other the reduction processes that use solid fuels.

Of the numerous and widely differing processes that exist, some of which are still in the development stage, only a few are described below; these are the processes that are recognized as being commercially workable and which have already proved their worth in industrial applications.

aa) Gas Reduction Processes

The HYL Process. Natural gas has for many years been used as a reducing medium for the production of sponge iron to direct reduction processes and has given every satisfaction in the large iron and steelworks of Hojalata y Lamina, Monterrey, Mexico; it takes its name from this works, being known as the "HYL" process. At present, this works produces some 800 tons of sponge iron per day. This process is taken as an example of gas reduction. The following schematic drawing (Fig. 7) briefly summarises the principle of this type of process, which is based on the reforming of natural gas to produce CO and H_2. The ore is charged into simple retorts which feature a low-grade refractory lining, the reaction temperature of the process being only slightly above 800 °C.

The cold primary gas leaving the reforming plant, composed mainly of hydrogen and carbon monoxide, is led through a reactor filled with hot sponge iron that has just come off the reduction phase; this heats the primary gas and cools the sponge iron. The carbon so formed in accordance with Boudouard's reaction remains in the sponge iron; in the following steel process, this carbon, on melting taking place, serves to reduce the remaining ferrous oxide (FeO) and supplies the necessary surplus carbon needed for intro-

ducing the boiling phase. The heated gas then flows through the
heat accumulator of the cooling reactor and then the reactor in
which the process is commenced, and initiates the reduction phase
in this reactor. It then flows through the secondary reactor and its
heat accumulator, and so on. Complete reduction of the ore takes
place in the primary reactor while preheating and a certain degree
of prereduction takes place in the secondary reactor. In other words,

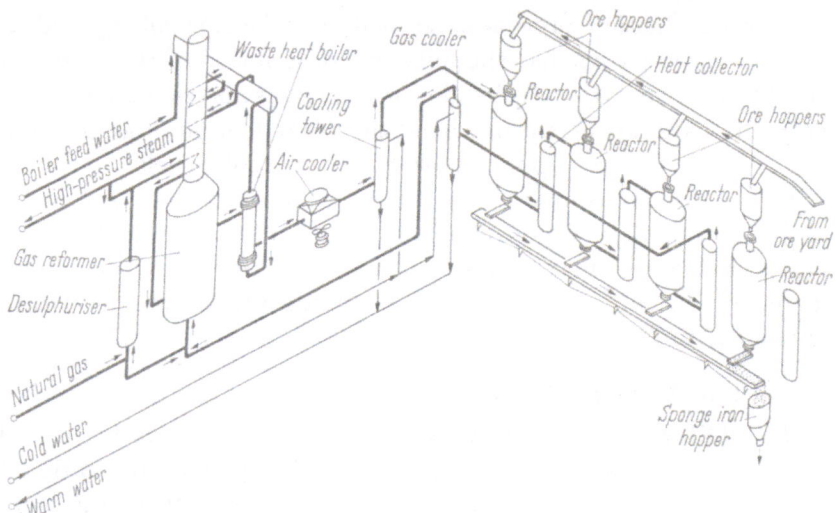

Fig. 7. Schematic representation of the HYL sponge iron process.

when 4 reactors are in use, a charge is placed in the first, the second
is preheated and prereduction takes place, complete reduction takes
place in the third, and cooling and removal of the finished product
in the fourth. The materials charged, whether lump ores, agglomerated or pelletised ores, naturally remain in the same reactor from
start to finish; the individual reactors merely go through the described cycle of operations.

Some 5 million kcal are consumed per ton of sponge iron produced to this process; this is a rather high figure, and consequently
the process can only be put to economical use in areas where natural
gas is available in large quantities and at a low price. Nevertheless,
the practical results obtained when using high percentages of sponge
iron made to this process have proved that its use in subsequent
steel processes, and in particular in electric arc furnaces, is both
feasible from the economic point of view and promising from the

technical point of view. No additions of solid or liquid fuels such as coal or oil are required by this process, all reduction and heating work being performed with gas alone.

The Purofer Process (see Fig. 8). It would at this juncture be appropriate to mention a shaft-furnace process which, although it is not yet in use on a major scale, has a promising future as regards industrial application on account of the fact that it is relatively simple to handle and uses orthodox metallurgical equipment.

This is the "Purofer" process, in which natural gas is heated to about 1,200 °C in a gas reformer by a regenerative system that at the same time acts as a catalyst, and reformed in a pre-mixing chamber to produce a mixture composed of hydrogen and carbon monoxide by adding oxygen or steam, or both, the temperature then dropping to roughly 1,000 °C. Less heat is required when reforming with oxygen than is the case when steam is used, whereas the latter is in itself less expensive. Gas is then conveyed through a second system that was heated up in the meantime. The catalyst itself is heated by the waste gas of the reduction furnace as such.

The ore (lump or pellets) is charged into the actual reduction furnace—which can incorporate many structural features of the classical blast furnace—through a

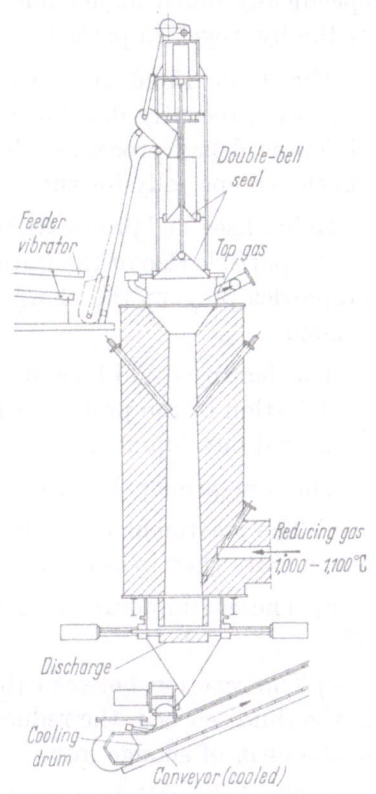

Fig. 8. Schematic representation of the Purofer process.

double bell-and-hopper arrangement and moves from top to bottom during reduction. The reformed gas has approximately 27% CO and 60% H_2.

A special facility removes the sponge iron at the bottom and throws it into a water-cooled drum (cooling is affected in a gas atmosphere). The reformed gas is introduced through the lower

3*

half of the shaft furnace. Virtually all operations can be automatically controlled. A sponge iron with approx. 96% Fe, more than 91% of this being metallic, can be produced from pellets with an Fe content of about 67%. The C content can be controlled in the range of 1 to 5%. As with the HYL process, the external structure remains unaltered.

Compared with a blast furnace of equal size, the throughput is specifically much higher due to the more intensive reducing effect of the hydrogen in particular.

The Esso Fluid Iron Ore Reduction Process (Fior Process). A process recently developed by Esso is described as an example of fluidised bed processes. This is the "Fior" process, which was developed specially for the smelting of ore fines.

In fluidised bed processes, sintering or pelletising is not necessary, as the particle size of the initial materials is such that their specific properties prevent the charge from rising or sinking, i.e. it is suspended.

This feature could be a great advantage for the later commercial exploitation of the process, which is at present in an early stage of industrial development.

The four principal features of the process are:

a) The treatment of finely divided solids in a fluidised state, in which state they behave as a liquid.

b) The manufacture of reducing gas from hydrocarbons (e.g. oil).

c) The reaction between the fine ore particles and reducing gas in the fluidised bed, the reduced Fe particles sinking to the bottom in the form of sponge iron.

d) The briquetting or pressing of the cooled sponge iron powder to firm lumps. There is no significant danger of back oxidation, particularly if suitable coatings or covers are used.

This and other sponge iron processes can be profitably linked with the electric arc furnace process. With continuous charging, the power consumption rates of the arc furnace not only compares favourable with those usual when melting scrap, but can under certain circumstances be even lower.

Fig. 9 shows a line diagram of the process.

bb) Direct Reduction Using Solid Fuels or Oil. It is obvious that reduction effects similar to those obtained using natural gas as described in the previous section, can be achieved by using solid fuels, possibly using oil or gas as additional sources of heat.

Several direct reduction processes using solid fuels have been developed along these lines. Some are retort processes, some shaft-furnace processes, and others rotary kiln processes. Of the many that exist, one known as the "SL-RN" process is taken as an example (Fig. 10).

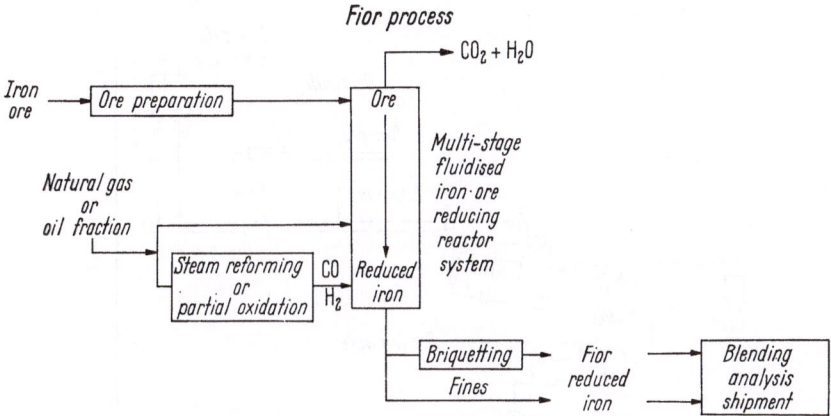

Fig. 9. Flow diagram of the Fior process.

It was developed by the Steel Company of Canada in cooperation with Lurgi, and also utilises the patents of the "RN" process developed by Republic Steel and National Lead. A medium-sized plant using this process has been giving good results for several years in Canada, producing some 150 tons per day, which means that the process can be regarded as having proved its industrial efficiency. This process uses lump ore, or fine ore below 10 mm which is sintered, or crushed and ground to a grain size suitable for pelletising, and rolled into pellets 12 to 15 mm in diameter; these contain certain percentages of fine coal for reduction purposes. Depending on the suitability of the ore, these pellets can be charged as baked; some ores will require 1 to 2% of bentonite for pelletising. The burden and additions needed for desulphurisation, together with any surplus carbon (coke, anthracite) needed, are then charged into the rotary kiln.

The kiln is heated by adjustable burners spaced along the entire length of the shell, plus a central burner located at the discharge end. As the kiln slowly revolves, the charge progresses from one end to the other, a distance of 40 to 100 m; during this journey it is heated up, and the oxygen in the ore is reduced to the highest possible degree by the carbon. As in the HYL process, the product discharged from the end of the kiln is sponge iron.

Here too, the operating temperature is so low that the ore mixture materials are not melted or sintered; this dependably avoids

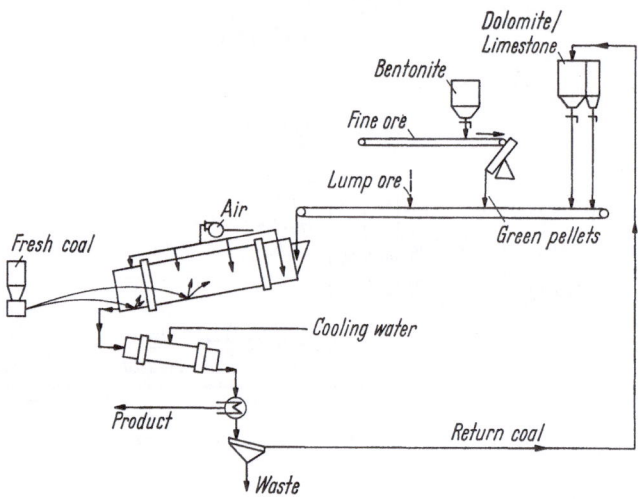

Fig. 10. Flow diagram of the SL/RN plant.

the formation of rings in the kiln. As opposed to the Krupp-Renn direct reduction process, the iron is not reduced to the molten state and withdrawn in balls, when the gangue is separated off by a magnetic process; instead, all the gangue remains in the ore or the pellet mixture. This means that from high-grade iron ore we get a sponge-iron pellet or sponge-iron ore with some 90% of metallic iron and a few percent of FeO; naturally, if the ore mixture contains more gangue, the iron content is that much lower, as the gangue remains in the pellet. This process is very economical from the heat consumption aspect, using some 3,500,000 kcal per ton of sponge iron; of this, about 20 to 40% can be obtained by burning gas or oil. The kiln temperature lies between 950 and 1,050 °C. Virtually any sort of coal can be used, including anthracitic coals or coals with a high content of volatile matter. If the latter type is used, it is possible so to control the process that no oil or gas need be fired. Combustion of the volatile matter provides the heat, and the solid part of the coal is the reducing medium. A schematic representation of this process is given in Fig. 10.

As all types of non-coking coal can be used, this process is extremely important to countries that have such coals but no natural or other gases at their disposal, and can form the basis of an iron and steel industry. The present stage of development in this field permits the building of kilns for up to 1,000 tons of sponge iron per day without any difficulty.

2. Prereduction

It is obvious that direct reduction processes can also be used for partial direct reduction, which is known as prereduction. The advantage of these prereduction processes is that they can be used in conjunction with one of the normal reduction processes, i.e. blast-furnace and electric reduction processes.

When direct reduction processes are used for prereduction only, two important changes are brought about. Firstly, they permit much higher outputs in comparison to a given amount of fully-reduced sponge iron, and secondly, the use of prereduced material in blast furnaces or electric reduction furnaces can substantially cut the required amounts of reducing and heating materials, i.e. coke in the blast furnace and electric power and reduction media in the electric furnace, and the capacities of these units can be considerably increased.

A rotary kiln process of this type can, of course, be profitably coupled with an electric reduction unit. In this context, and in keeping with the present state of development in this sector, at least three processes that can work in this way are worthy of mention. Firstly, there is the Strategic Udy process, developed by Koppers, USA; secondly, the Elkem prereduction process, developed by Elektrokemisk, Sweden; and thirdly, the SL-RN process with the standard low-shaft electric furnace. This latter process is described below as an example.

Depending on the Fe content of the charged ore, a very high degree of prereduction can be obtained using this process, possibly between 50 and 80%. When prereduced material (lump ore or green pellets) is charged into a standard half-open or open low-shaft

electric furnace, power consumption can be radically cut. For
example, when smelting prereduced iron ores with only some 44%
Fe in standard low-shaft electric furnaces, power consumption rates
of only 1,300 kWh/t have been achieved using cold charges, and it
is obvious that if hot charges are used, power consumption can be
reduced by several hundred kWh. On the other hand, the kWh
consumption per ton of hot metal when using hot charges should
not be estimated too low, especially where half-open furnaces are
concerned.

This high degree of prereduction means that the volume of gas
produced in the electric reduction furnace is very low; conse-

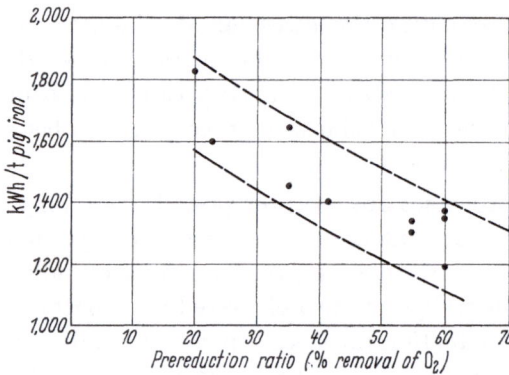

Fig. 11. The influence of
the prereduction ratio (%
removal of O_2) on power
consumption (electric low-
shaft furnace).

Values ascertained during
trial operations.

quently, inexpensive, easily operated low-shaft electric furnaces
of open design can be used.

Similar processes exist that use sintering belts instead of rotary
kilns, such as the DLM (Dwight-Lloyd-McWane) process.

Fig. 11 shows a certain range of reductions in power consumption
that can be attained with an increasing degree of prereduction.
The most important feature of this process is that the low power
consumption rates make it very advantageous for works paying
substantially more than about 6 mils/kWh (2.5 Dpf). A further
advantage that should be emphasised is the fact that the low-shaft
electric furnaces used in conjunction with this process have roughly
twice the capacity of low-shaft electric furnaces used in standard
processes. The future will show that the use of these processes will
put the electric pig iron process on an entirely new footing, especially
in countries that have no coking coal or natural gas at their disposal.

Coke consumption in conventional blast furnace operations can also be cut if prereduced stock is used. Numerous series of investigations have already been carried out in this respect during which the coke consumption rates were reduced to below 400 kg/t hot metal.

To complete the picture, mention is made of the fact that in countries that have generous natural gas resources, and possibly low-grade ores that do not lend themselves to dressing, the natural gas prereduction process on the HYL or Purofer system can be used in conjunction with conventional low-shaft electric furnace processes or the blast furnace process.

3. Blast Furnaces

Most of the world's pig iron is still produced by smelting ores in the blast furnace. The processes used for making pig iron outside the blast furnace, although they have in the meantime given proof of their efficiency, still play but a minor role, or are adapted to special conditions for low-grade ores or fuels other than coke.

Blast furnaces have greatly increased in size during the past few years. In 1960, the average output of German blast furnaces was 700 t/24 h; today, a hearth diameter of 9 m plus is the standard taken in Western Europe for new blast furnace plant. Blast furnaces with such hearth diameters have capacities of 1,000 to 2,000 t/24 h, and the development of furnaces with capacities of 5,000 tons per day is not far off. These high capacities were made possible only by the change-over to burdens made up of high-grade ores and with virtually uniform physical compositions, this being the only way to guarantee uniform gas distribution throughout the column of blast furnace stock and a high degree of indirect reduction.

In order to obtain a uniform physical composition, lump ores are crushed to a uniform lump size, nowadays in general between 10 and 30 mm, and the fines are agglomerated by pelletising or sintering. With modern blast furnaces, the once standard practice of charging run-of-mine ore has completely disappeared. Quite naturally, the coke is adapted to suit the burden, 20/40 or 40/60 mm now being the usual lump size for blast furnace coke.

A rich ore burden, a high burden yield, and a low slag volume can, of course, permit a lower coke rate. At a net burden yield of 60%, coke rates of 600 kg/t blast furnace metal and less can easily be attained.

Slag volumes can be reduced to less than 300 kg/t blast furnace metal. Leading blast furnace operators have misgivings about extremely low slag volumes, in view of insufficient desulphurisation, and it would therefore appear that certain limits are set here.

During the past few years, experience has been gathered all over the world in the injection of oil or natural gas into blast-furnace hearths. The present stage of development can be summarised as follows:

Fuel oil injection—and the same applies to natural gas as appropriate—has its limit at some 100 kg oil/t blast furnace metal. The optimum value is probably around 50 kg/t blast furnace metal, the "substitution ratio" reaching its maximum at this point, i.e. when this quantity of oil is injected, about 1.3 to 1.5 kg of coke are replaced by 1 kg of oil, whereas at 100 kg of oil the ratio drops to roughly 1. This also clearly defines the profitability of oil injection. For example, if coke costs 25 $/t (100,— DM) free furnace, or in other words 0.025 $/kg (0.10 DM), and the oil plus handling costs (pumping, heating, injecting etc.) 20 $/t (80.— DM) or 0.02 $/kg (0.08 DM), the substitution ratio would be 16:10. In other words, if 50 kg of oil are injected, $50 \times (0.0325 - 0.0200) = 0.625$ $/t (2.50 DM) blast furnace metal are saved by oil injection.

Where prices for coke and oil differ from those quoted above, this saving will, of course, be higher or lower.

The furnace profile was at one time a major factor as regards blast furnace design. Present-day high throughput figures, complete physical preparation of the burden, and the high burden yield have led to a great simplification of the furnace profile and reduction of the "batter", i.e., the stack angle from the throat downwards to the bosh was increased, and the blast furnace so given a slim profile. In the USA and the USSR, blast furnaces are being pressurised on an increasing scale; this must be given due consideration when new blast furnaces are being planned. The pressurised furnace has better gas distribution at reduced gas velocities, and reduced flue dust volumes.

Charging installation selection and design is largely determined by the burden to be used; most blast furnaces now feature inclined hoists, whether bucket or skip-charging is used. Buckets can also be used with vertical hoists, which are often necessitated by cramped space conditions; this applies to small and medium-sized blast furnace plants, in particular. Modern high-capacity blast furnaces, on the other hand, nearly always feature skip charging, as this system is suited to increasing furnace capacities, whereas lack of space often makes the adaptation of an existing vertical hoist to increased furnace sizes extremely difficult.

In addition to skip charging, *conveyor belt charging systems* have been successfully introduced in recent years.

Using a belt charging system, one man can control and supervise the charging of a group of two or three blast furnaces from a central control point. The first installations of this type have been successfully taken into use for the first time ever in the Dillingen (Saarland) and USINOR iron and steelworks, and in the new French iron and steelworks at Dunkirk. Belt charging systems also permit the use of computer control.

The *bell-and-hopper system* must be designed to keep gas losses as low as possible. This applies to the space between the two bells *and* the need for a gas-tight seal; the latter point is of great importance in view of the trend towards pressurised furnaces as discussed above.

Top gas evacuation is no longer effected centrally, but through two, preferably four, laterally-arranged downcomers.

Blast-furnace cooling systems have been developed along various lines. The spraying of blast furnace shells is still a usual practice, but some very successful trials have been carried out with circulating cooling systems. The bosh and more or less extensive parts of the stack are still cooled by cooling boxes, which are nowadays arranged in a checkerwork pattern and welded into the shell, which in the case of modern blast furnaces of all sizes often extends up to the throat.

Stack cooling assumes more importance when smelting low-grade ores, as the throughput of burden and consequent wear of the furnace lining increase. The "Burgers" furnace with steel-plate shell and spray cooling system no doubt has certain advantages as regards ease of repair, but has the disadvantage of higher cooling losses. *Cooling water supplies* are discussed later.

The temperature of the cooling water running down the shell is generally kept below 30 or 40 °C. If the water used is not extremely hard, temperatures of up to 50 °C can be used without any fear of serious trouble through boiler scale deposits.

The linings of blast furnaces used to be 950 mm thick, but now never exceed 650 mm. With linings of this thickness, the cooling boxes also extend through to the shaft profile. Taken together with the now standard checkerwork arrangement of the cooling boxes, this provides a substantially improved anchorage for the lining.

With few exceptions, the hearths and boshes of modern blast furnaces are lined with carbon blocks. Rammed carbon compounds, frequently used as an emergency measure during the war, have not proved satisfactory as the compound tends to crack or dissolve in the hot metal. In earlier days, the use of very small carbon bricks in the hearth was recommended from many quarters, but this too has been dropped again in the meantime; the practice now is to use the largest blocks practicable, in the region of $400 \times 600 \times 600$ mm. The use of well-ground blocks with close tolerances produces a virtually joint-free hearth lining without any difficulties being experienced.

Blast tuyères, which are still preferably made of copper, and recently of dependable aluminium alloys, have not undergone any radical changes in design during the past few years. Their design must permit rapid installation and dismantling, e.g. by means of a special block and tackle running around the blast furnace. The possibility of checking the blast volume of each tuyère by means of simple differential pressure measurements is generally known but seldom used.

Various types of pneumatic, steam, and electric-powered *taphole guns* are available on the market; each type has its advantages and disadvantages, and no order of superiority can be given.

Blast. Modern blast-furnace operations demand a constant *blast volume* instead of the constant *blast pressure* which was the sole requirement in earlier days. This means that a pressure reserve has to be maintained to cater for any sluggish working of the furnace. Blast pressures of 0.7 to above 1.5 atmospheres gauge are therefore normal today for medium-sized and large blast furnaces, and the blowers are accordingly designed for a delivery pressure of 1 to 2 atmospheres.

The *blast temperature* can nowadays be kept up to 1,000 °C without difficulty, and the actual temperature required depends on the type of ore burden to be smelted. For example, the smelting of Minette ores and siderite from the Styrian "Erzberg" (ore mountain) can only be effected smoothly at blast temperatures of 400 to 550 °C. At higher temperatures, scaffolding is likely to occur. The pasty zone of Styrian siderite is rather broad; consequently, excessive blast temperatures cause bridging in this zone of the shaft.

Today, nearly all blast furnace plants use *Cowper* stoves to heat the blast, but since the development of steel recuperators some plants have used these with success.

Hot-blast stoves have for some years shared a more or less standard design; the checkerwork, their surfaces and dimensions are also sufficiently well known, and nothing substantial can be said regarding the design and selection of these facilities[1].

It need not be specially emphasized that the Cowper stoves have their own blowers, that temperature control equipment is used, or that semi-automatic or fully-automatic switching systems are used. Many people prefer the semi-automatic control system in which the operator has to initiate switching — instead of the temperature controller or a clock in fully-automatic systems — as this better ensures supervision of the hot-blast stoves by the operator.

It is advisable to thoroughly check the question of Cowpers vs. steel recuperators when planning small blast furnaces.

Blast-furnace Blowers. The various types of *blowers* are frequently the subject of detailed discussions among specialists. Today, we can choose from:

turbo blowers driven by steam, electricity or gas, and
reciprocating blowers driven by steam or gas.

Fundamentally, reciprocating blowers are more suitable for overcoming greater pressures, and consequently it is easier to maintain the required (constant) blast volume with reciprocating blowers when furnace resistance is high. However, the same requirements can be met by appropriately designed turbo blowers, and they can be much better regulated. Hence, they are nowadays usually preferred to reciprocating compressors, particularly those of the gas type.

The pulsations caused by reciprocating blowers constitute defect sources for measuring and control equipment, greatly impede measurements using sharp-edged orifices, and frequently make these measurements impossible.

With *steam and electric drives*, the appertaining boiler plant or power generating equipment must be taken consideration during

---

[1] Stoves have been developed for ultra-hot blasts up to 1,200 °C. They feature silicon linings and have separate combustion chambers.

the planning stage; if gas blowers are used, this extra plant and equipment is not required.

Table 4. *Comparison between standard blast-furnace blowers*
(according to PFENNIGER Loc. cit.)

Description	Unit	Reciprocating blowers (Gas)	Steam	Gas	Electric
				Turbo blowers	
Heat consumption per 1 atmosphere gauge blast pressure	kcal/Nm³	160—200[a]	185	137[a]	176
Thermal efficiency	%	16.4	15.9	21.5	16.7
Blower weight per 1,000 Nm³/h	t	11	—	2	—

[a] Without waste-heat boiler.

The gas turbo blower deserves special mention[1]. Fig. 12 shows a schematic representation of a blast-furnace turbine installation at the Düdelingen works of Arbed.

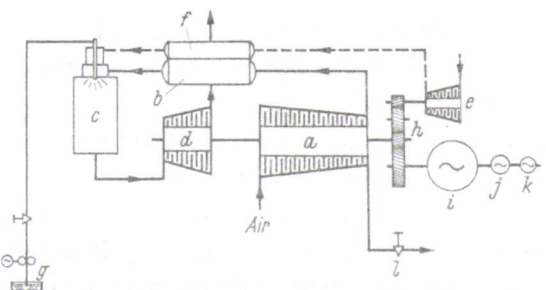

Fig. 12. Diagram of a 5440 kW gas turbine plant (BBC).
a axial compressor; *b* air preheater; *c* combustion chamber; *d* gas turbine; *e* blast-furnace gas compressor; *f* blast-furnace gas preheater; *g* fuel oil pump; *h* gearing; *i* generator; *j* starter motor; *k* exciter; *l* air delivery — 16 m³/sec at 2.8 atm abs. for the steelworks, or 22 m³/sec at 1.5 atm abs for the blast furnaces.

The difference between this installation and earlier facilities is that a generator is so arranged that it is subjected to a heavier load when the blast is throttled, and runs under virtually no-load con-

---

[1] KARBER, W.: Gasturbinen und ihre Anwendungsgebiete (Gas Turbines and their Applications). Stahl u. Eisen 72 (1952) pp. 885/92. — H. PFENNINGER: Betriebserfahrungen mit Gasturbinenanlagen in Hüttenwerken (Operating Experiences with Gas Turbine Plant in Iron and Steelworks). MTZ 16 (1955) p. 47 ff. and p. 193 ff.

ditions when the blast is full on. This means that the gas turbine practically always runs under an almost uniform optimum load. The thermal efficiency is given at 21.5%.

The *blower boiler house* for steam-drive blowers is nearly always equipped with gas boilers. A steam pressure of LP = 40 atmospheres gauge would appear to be most suitable. Oil firing for standby duty is expedient, and, as it is very seldom needed, acceptable. If coal were to be used, the furnace would have to be of special design that would be detrimental to the thermal efficiency of the normal gas firing operation.

The *blowing equipment* for all blast furnaces or each group of blast furnaces (8 and more blast furnaces are best divided into groups of 3 to 5 furnaces) is concentrated in one *blower house*. If at all possible, the blast lines should be laid in such a way that each furnace can operate on its own line, and all blowers can deliver together to all furnaces. Finally, it must be possible to switch the reserve blower to any furnace *and* to the common line. One reserve blower per 4 to 5 blowers should be adequate, 6 blowers may need 2 reserve units, and so on.

Gas Cleaning. Basic gas cleaning facilities comprise:
Wet scrubbing plant (Theisen, Zschokke, Dingler, etc.)
Dry filter plant (Halberg-Beth)[1]
Electrostatic precipitation plant (Lurgi-Elga, etc.).

In actual fact, combinations of two methods are just as often used today. These installations feature primary coolers, gas cyclones, and other ancillary equipment.

However, in all cases where blast-furnace gas has to be cleaned to "mechanical purity", i.e. where the maximum permissible dust content per Nm^3 of dry gas is 0.02 g, cooling and consequent "drying" of the gas—appropriate reduction of the steam content—is also imperative.

The same applies when blast-furnace gas has to be conveyed through very long lines. Water condensation—which occurs to a pronounced degree during the cold season—can lead to breakdowns; in bad cases, the gas line practically freezes solid.

In all such instances, the hot gas coming from the blast furnace—carrying from 50 to over 100 g/steam per Nm^3—must be cooled to $20-30$ °C somewhere in the gas cleaning plant. If the gas pressure

---

[1] See note 2 on page 48.

at the blast-furnace top is insufficient for conveying the gas through the cleaning plant and the following pipework, conveyance of the gas is best combined with the gas cleaning plant; the Theisen gas washer, for example, combines the two in a single unit in an ideal fashion. The gas blower is normally arranged on the clean gas side of the plant in order to prevent contamination and subsequent stoppages. Great care must be taken to ensure that *no partial vacuums* occur anywhere in the gas cleaning system; in electrostatic precipitation facilities the results could be disastrous. In all other types of gas cleaning plant, too, partial vacuums and the air so admitted into the system, together with the pyrophoric flue dust, can lead to fires breaking out in the cleaning plant[1].

The following table compares the investment costs of the two most important systems[2] (complete plants, including cooling water facilities).

	Wet			Electrostatic		
	Cleaning Systems					
Nominal rating Nm³/h	80,000	180,000	250,000	200,000	250,000	600,000
Investment costs million $	1.0	1.8	2.25	1.5	1.8	4.0
Investment costs per 1000 Nm³/h $ (1966)	12,500	10,000	9,000	7,500	7,250	6,650

Blast and Gas Pipes, Gasometers. The fact that all pipework for liquid and gaseous media must be of the correct aerodynamic design, particularly with regard to the selection of the cross-sections and arrangement of elbows, branches, and slide valves, is self-apparent and need not be covered in detail.

Ring mains are commonly used today, as they offer great flexibility should defects occur in any particular section.

The need to build *gasometers for blast furnace gas* is often over-estimated. A blast furnace plant with an annual output of 1,000,000 t produces 400,000 to 500,000 Nm³ of blast furnace gas

---

[1] See F. Lüth: Richtlinien für den Wirtschaftlichkeitsvergleich von Hochofengas-Reinigungen. Arch. Eisenhüttenw. 5 (1931/32) pp. 223/30 (Guides for Comparison the Economic Aspects of Blast Furnace Gas Cleaning Plants).

[2] Virtually no dry cleaning plants are built these days.

per hour; a gasometer with a capacity[1] of 100,000 m³ would in this instance only be able to cater for the gas produced in 12 to 15 minutes. This cannot be looked upon as a gas storage system; the gasometer merely acts as a pressure equalizer, a function that can be carried out just as well and much more cheaply by modern control appliances. This is confirmed by experience gathered during World War II; in Germany at least, civil defence regulations forbade the use of gasometers, and the various works just had to manage without them. This certainly hampered the management of the gas supply system, but no iron and steelworks ever experienced any serious difficulties when operating without gasometers.

Measuring and control systems are nowadays indispensable for correct and smooth blast furnace operations. Stock line recorders, blast pressure and volume recorders at the furnace itself, temperature and volume control in the blast heating installation, semi or fully automatic-switching equipment for the blast heating facilities, and measuring and control equipment in the blower house are vitally necessary to a modern blast furnace plant.

In addition to measuring and control facilities at the various consumer points, a centralised control system for all forms of energy is not merely expedient, but imperative in integrated iron and steelworks, where blast furnace gas is often the mainstay of the steelworks and rolling mill fuel systems. The most important measured values of the main forms of energy — blast-furnace gas, coke-oven gas, electric power, water, etc. — are recorded in a central control room. From here, fluctuations in production and consumption are balanced out. This applies not only to the fluctuations occuring during the three daily shifts, but particularly to fluctuations arising over the weekend.

Pig casting machines are very advantageous for dealing with hot metal produced on Sundays and holidays, as well as for the emergency casting of hot metal from blast furnaces and other furnaces that produce metal continuously and therefore have either no pig beds at all or only inadequate facilities. In addition, the demand for machine-cast foundry pig iron is nowadays very great, as these pigs are free of sand and offer many advantages in foundry use. The casting machine can be located to one side of and at a consider-

---

[1] Investment costs: 5 to 10 $ per m³ (20 to 40 DM) in the case of high-capacity gasometers (Basis 1965).

able distance from the blast furnace plant. The operating costs of a pig casting plant amount to some 2.50 $/t (10.— DM) pig iron (Basis 1965).

Slag utilisation must be given very careful consideration during the planning stage. The quantity and quality of the slag depend in large measure on the type of burden and the smelting process. For example, basic pig iron burdens, and in particular foundry iron burdens made up of Swedish and other high-grade foreign ores, and basic Minette ores, produce a slag that can be used as a building material, for the manufacture of cement, or as a foamed slag and so on. As opposed to this, the slag obtained in "acid" smelting operations cannot be used at all unless the slag ratio is higher than $p = CaO/SiO_2 = 0.75$, and even then it is *unsuitable* for cement or foamed slag, and needs very careful treatment and processing in any case. Nevertheless, good paving bricks, road ballast, small stones and so on can then be obtained. Slag sand can easily be made by granulating and used for the manufacture of numerous items such as building brick, etc.

If slag cannot be sold at any particular moment and must be placed on a dump, no credit balance is forthcoming; instead, additional transport costs have to be met. These can quickly amount to 0.25—0.75 $ (1.— to 3.— DM) and more per ton of pig iron.

In other words, the slag problem must be closely scrutinised well in advance during the planning of all blast furnace plant; if lean ores are to be smelted, the amounts of slag accumulating will be much greater and the need to make certain of a definite slag credit so much the more imperative. If this aspect is neglected, the absence of a credit for the slag and the extra costs incurred for transportation to the slag heaps could seriously affect or even jeopardise the economical operation of the blast furnace plant.

Oxygen. The use of an oxygen-enriched blast has progressed beyond the trials stage[1]. Operational experience in Germany indicates that the optimum oxygen content lies between 25 and 28%, at which value furnace conditions and output were improved and coke consumption reduced. A further increase in the oxygen content proved unsatisfactory, as the gas volume dropped excessively.

---

[1] LENNINGS, W.: Sauerstoffangereicherter Gebläsewind im Hochofenbetrieb (Oxygen-enriched Blast in Blast Furnace Operations). Stahl u. Eisen 55 (1935) pp. 533/44 and 565/72.

Blast Furnace Operations. Up to a few years ago the "*n*-formula" was still used for planning purposes, i.e. in plants with up to six blast furnaces, only $n - 1 = 5$ furnaces, for example, were allowed to be in operation at any one time.

In plants with more than 6 blast furnaces, two (formula: $n - 2$) had to be kept in reserve at all times. Nowadays, the use of all available furnaces, a long-standing practice in the USA, and one that has proved its efficiency, is gaining ground.

A simple calculation demonstrates the profitability of this modus operandi. If a campaign of 8 to 10 years is taken as a basis—this depending on the burden used—and assuming that the blast furnaces are duly shut down and relined at the end of this campaign, the following calculation can be made as an example for a blast furnace plant with five 700-ton furnaces (see page 52).

The calculation shows that it costs much more to keep a stand-by furnace ready for use at all times than it does to run the entire plant at a (time) utilisation factor of 96%.

The development trend is towards ever-increasing furnace capacities. Distribution of the pig iron capacities among the lowest possible number of the largest possible furnaces produces a substantial drop in investment costs, space requirements, and operating costs; however, this is all subject to a rigid time table being adhered to as regards campaigns, relining dates, and relining schedules.

Blast-furnace ratings are still calculated on the load imposed on the furnace hearth by the combusted coke in terms of kg of coke per m² per hour. Normally-driven blast furnaces have hearth loads of between 800 and 1,000 kg/m²h. Really hard-driven furnaces have higher values. Blast furnaces operated on a burden of lean ores with 30% and less Fe are generally charged with unscreened run-of-mine ore, or with alternate charges of small ore and lump ore. Experience has shown that blast furnace operations are smoother and more economical at lower hearth loads, i.e. about 600 kg/m²h and less. Of course, the output of a blast furnace in tons of iron per day increases as the coke rate (kg/t pig iron) is reduced, working on the basis of an uniform hearth load. Fig. 13 shows the relationship between the output of pig iron and the hearth diameter at various coke rates between 500 and 800 kg/t pig iron, taking a hearth load of 1,000 kg/m²h as a basis. With lower hearth loads, the iron output is reduced in ratio. To illustrate this aspect, a second ordinate scale is shown for a hearth load of 750 kg/m²h.

4*

		Operation with	
		4	5
		blast furnaces	
Campaign duration	Years	8	
Relining time	Years	1/4	
Utilisation period per blast furnace		$4/5 = 0.80$	$8/8.25 = 0.97$
Utilisation period per calender year	Furnace years	$5 \times 0.8 = 4$	$5 \times 0.97 = 4.85$
Capacity of one blast furnace (700×365)	t/year	250,000	250,000
Capacity of the blast furnace plant[a]	t/year	1,000,000	1,200,000
Investment costs per blast furnace[b]	DM	30,000,000	
	$	7,500,000	
Investment costs of the blast furnace plant	DM	150,000,000	
	$	37,500,000	
Capital service [c]	DM/year	10,500,000	10,500,000
	$/year	2,625,000	2,625,000
Maintenance of one blast furnace[d]	DM/year	1,000,000	—
	$/year	250,000	—
Total capital service and costs of keeping one blast furnace ready for use at all times	DM/year	11,500,000	10,500,000
	$/year	2,875,000	2,625,000
ditto per ton of pig iron (capacity)	DM/t pig iron	11,50	8,75
	$/t p.i.	2,80	2,19

[a] $4 \times 250,000$ or $4.85 \times 250,000$.

[b] 250,00 t/p.a $\times$ 30 $ (120 DM) per t p.a. capacity.

[c] According to the SPITZER tables, 30 years' depreciation at an interest rate of 7% produces an average service on capital of 7%.

[d] The costs of keeping a 700-ton furnace ready for operation at all times are estimated at 10% of the processing costs for normal operations, i.e. at approx. 250,000 $ (1 million DM) p.a.

4. Low-shaft Furnaces

The low-shaft furnace was developed to permit the smelting of ores to produce a serviceable pig iron using fuels other than high-grade metallurgical coke. Various processes have been developed for this application, each using a different method to attain the same goal. However, the division of these processes into two main groups, depending on the use to which the fuel is put, would appear to be both expedient and correct[1]; on the one hand we ought to think in terms of a "small blast furnace" when smelting is carried out using classified coke—as in the Ko-We and Dingler processes—or in terms of a "gas producer with a hot-metal taphole and a slag notch" when crude coal (regardless of origin) is gasified and ore smelted at the same time. In the first instance, where well-classified ore and coke are charged in the smallest lump sizes possible, but with no dust whatsoever, extremely uniform furnace working is achieved; the rising gases are well distributed, the reduction conditions good, and the top temperature reduced, so permitting a lower stock-line level than usual. If a furnace of this type is always charged with these burdens, the height of

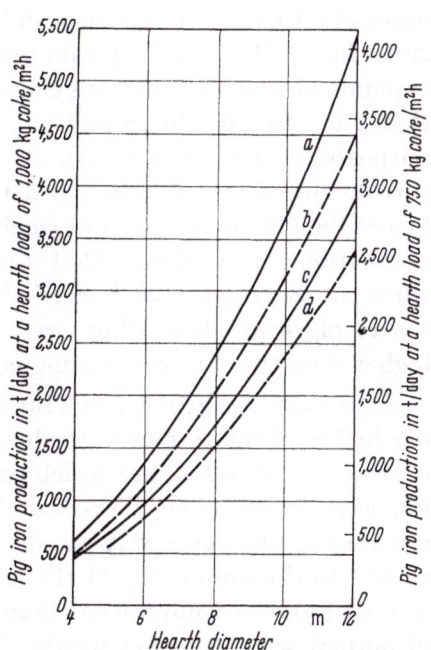

Fig. 13. Blast furnace capacity in tons pig iron per day as a function of the coke rate and hearth loading.

a coke rate 500 kg per ton pig iron; b coke rate 600 kg per ton pig iron; c coke rate 700 kg per ton pig iron; d coke rate 800 kg per ton pig iron.

[1] See F. LÜTH: Die Wertigkeit verschiedener Brennstoffe bei der Verhüttung im Hochofen und anderen Schachtöfen. Stahl u. Eisen 76 (1956) pp. 317/22. (The Values of Various Fuels in Blast Furnaces and other Shaft Furnaces), — E. E. HOFMANN: Stahl u. Eisen 74 (1954) pp. 1464/68. — M. PASCHKE: Die restlose Vergasung von Gaskohlen über Eisenkoks in Schachtöfen; ihre Bedeutung für die Gas- und Hüttenindustrie. In: Internationale Tagung über die restlose Vergasung von geförderter Kohle. Institut National de l'Industrie Carbonière, Liège. Ixelles-Bruxelles (1954) pp. 211/19 and 230/32 (The Complete Gasification of Coal above Iron Coke in Shaft Furnaces; its Significance for the Gas and Metallurgical Industries). — H. LINDE, K. SCHWINDT and M. PASCHKE: Stahl u. Eisen 75 (1955) pp. 691/93 (Hochofenaussch. 287).

the stack can be reduced and the furnace can with a certain degree
of justice be called a "low-shaft furnace".

Where coal is used for smelting purposes, the tarry gases rising
up the stack make a minimum top temperature of 500 °C or higher
necessary to prevent tar deposits in the stack section above the
stock line. These tar deposits are always mixed with fairly large
amounts of flue dust and are consequently more difficult to handle
than pure tar. At the same time, the usability of the tar for other
purposes is seriously affected. The obvious thing to do is to burn
this tar and dust-laden gas in a boiler plant immediately after it
leaves the gas-producing and smelting furnace. However, detailed
calculations have shown that the process is uneconomical even
when the tarry gases are burned. The carbon consumption rate per
ton of pig iron when using coal as a smelting fuel is many times
higher than the rate when using coke[1].

In any event, the low-shaft furnace process using coal as a smelting fuel is of interest only to those countries that have no coking
coal at their disposal and which can manage with furnaces of these
low capacities. If the low-shaft furnace is charged with coke, it
operates in the same way as any other small blast furnace, and,
thanks to the uniformity of the burden, it supplies a pig iron with
a more uniform composition than any produced in blast furnaces
of normal size. In other words, the small blast furnace will be of
importance wherever small amounts of special-grade pig iron are
required.

5. Electric Pig Iron Furnaces

The electric reduction furnace, successfully introduced many
years ago for the production of carbide and ferroalloys, was used
as a basis for the development of the pig iron furnace as long ago
as the thirties. As long as the gases were allowed to escape into the
atmosphere, however, the electric pig iron furnace could not compete
at all with orthodox blast furnaces; its use was therefore restricted
to a limited number of special cases where very cheap power was
available from hydro-electric power stations and where at the same
time expensive imported coke was the only fuel available. These
conditions have undergone a radical change over the past few
years, and the electric furnace is now an economical and therefore

---

[1] See footnote on page 53.

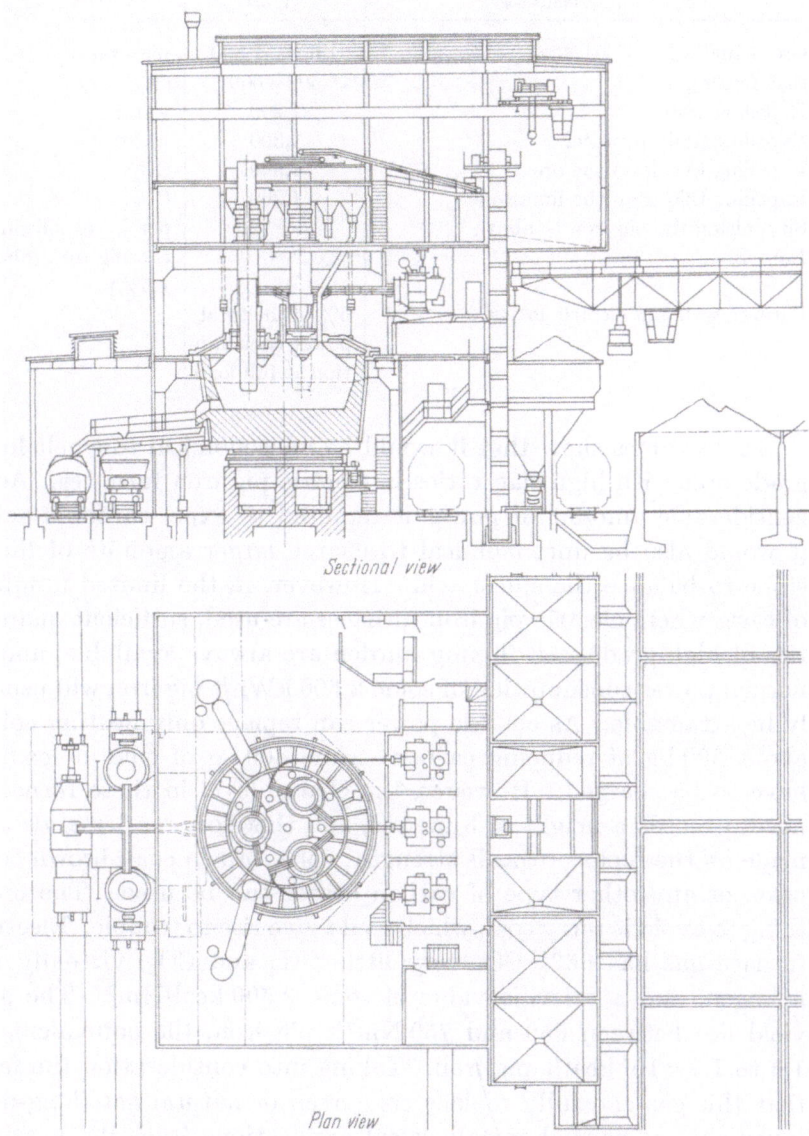

Sectional view

Plan view

Fig. 14. Diagram of an electric reduction furnace.

attractive proposition in many cases, even in Western Europe, especially in view of the fact that the capacity of this furnace has been increased to 300 t/day. The following table gives the characteristic power consumption values for electric smelting operations:

Power requirement for:	kWh/t	per 1 ton of
Ore reduction	1,600—1,800	pig iron
Slag forming	500—600	slag
Expelling water	1,200	water
Expelling hydrate water	1,300	water
Expelling CO_2 from the ore	1,500	CO_2
Expelling CO_2 from the limestone	1,200	CO_2
Siliconising the pig iron to above 1.5% Si	12	per kg of silicon in the pig iron (above 1.5%)
Furnace wall and electric losses	5% of the total power consumption at full load	

These values show that it would be uneconomical to smelt low-grade ores with high slag ratios in electric pig iron furnaces. As a considerable amount of power is required to expel carbonic acid, it would also be uneconomical to charge larger amounts of limestone to balance the silicic acid. However, in the limited number of cases where electric pig iron furnaces are used, sufficient quantities of high-grade, self-fluxing burden are always available, and a normal power consumption of some 2,500 kWh/t pig iron will usually be attainable. As electric power can replace only heating coke, about 300 kg of reducing carbon—some 400 kg of coke or coal—have to be charged. However, the charge level in these furnaces never exceeds a height of 3 metres, and therefore no demands are made on the fuel as regards strength; coke, brown coal, brown coal coke, or any other type of carbon bearer can be used. The only thing to avoid is tarry coal, which could give rise to trouble. Electric furnace gas has 65% CO, very little CH_4 and CO_2, virtually no nitrogen, and a calorific value of some 2,500 kcal/Nm³. The gas yield lies between 650 and 750 Nm³/t pig iron, the equivalent of 1.6 to 1.9×10^6 kcal/t pig iron. Taking into consideration the fact that this gas can fully replace coke-oven or natural gas (long-distance gas) in normal metallurgical applications from the heating point of view, it may justly be described as being the equal of long-distance gas in this respect. Normal blast furnace gas yields about 4×10^6 kcal per ton of coke, but some 50% must be deducted for losses and blast-furnace consumption, the remaining 2×10^6 (approx.) being valued at about half the present price for coke-oven or natural gas, which means that the gas credit per ton of electric furnace pig iron is in fact 1.7 times higher. As an example, Fig. 15 shows the acceptable power price for an electric pig iron furnace, plotted as a function of the price for blast-furnace coke, subject to certain conditions (*a*), and the equivalent price of power when taking the service on capital into consideration (*b*), which in the case of an electric pig iron furnace is about half that applicable to blast furnace plant.

As this illustration of a standard, modern electric pig iron furnace shows, these units are of relatively simple design and easy to

Fig. 15. Equivalent power price "X" for electric reduction furnaces as a function of coke price "P".
Basis: Blast furnace with 1000 kg coke at "P" DM-$/t, 2000 Nm³ blast-furnace gas credit at 2.50 $-10 DM/1000 Nm³, service of capital 7.5% of 40 $-160 DM/t/y.
Electric reduction furnace with 400 kg reducing fuel/t pig iron at $0.8 \times P$ $-DM/t, 2500 kW/t pig iron at "x" cent-pfennig, 1.75×10^6 kcal electric furnace gas/t pig iron at 5 $-20 DM/10⁶ kcal, service of capital 7.5% of 20 $-80 DM/t/y.
Lines *a* and *b* are based on the values, Line *a* without, line *b* including service of capital.

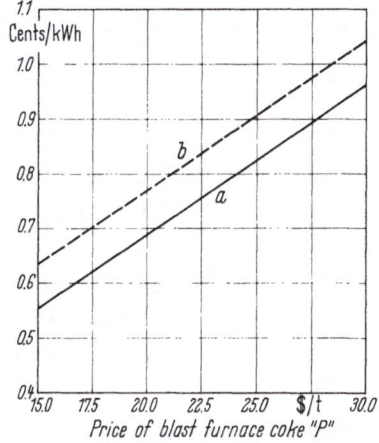

operate. The required heat is generated by the electric resistance of the charge.

The present state of development in this field permits the building of furnaces for some 300 t/day. Electric pig iron furnaces incur none or only a minor proportion of the investment costs needed in connection with blast furnaces for blast heating and compressing plant, for the cleaning of large volumes of gas, and for water cooling systems.

In spite of this, the investment costs for high-capacity blast furnaces per ton of pig iron ore are, of course, much lower than those required for electric pig iron furnaces of substantially lower capacity. The question as to which type of furnace is more suitable for a particular set of circumstances must be carefully examined from case to case; the operating costs and service on capital must be taken into consideration before a decision is reached.

Fig. 16 shows power consumption curves as a function of the Fe content of the ore burden.

It is stressed that the electric pig iron furnace is not subject to the current peaks often feared in electric arc furnaces. It operates very uniformly in this respect and consequently has no adverse effects on the work's power mains or the grid system.

In addition, if prereduced burdens or hot charges are used, the power consumption rates can be substantially reduced. Fig. 11 (page 40) shows power consumption figures for a given burden as a function of the degree of prereduction; power consumption figures of about 1,300 kWh/t for relatively lean burdens are technically well-founded.

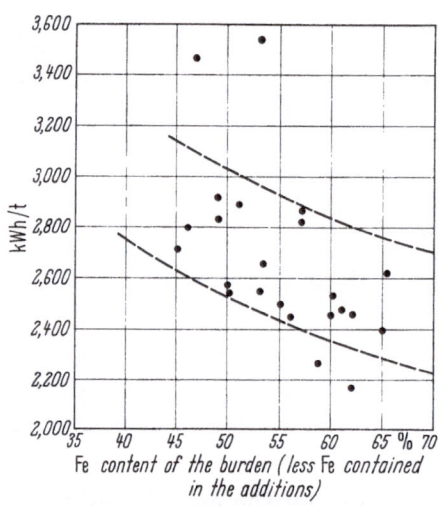

Fig. 16. Power consumption as a function of the Fe content of the burden (less Fe contained in the additions).

Values ascertained during trial operations.

This makes the electric pig iron furnace attractive under conditions that would earlier have incurred excessive costs; this becomes even more apparent when the following facts are taken into consideration:

The use of prereduced burden or hot charges not only reduces the specific power costs, but also increases the furnace output. It may be said that a furnace producing some 250 t/day from normal charges can produce 500 t/day under certain conditions when charged with material that has been prereduced to a high degree. However, increased furnace production is in this case accompanied by an increase in investment costs for the prereduction facilities. On the other hand, the use of prereduced burden permits the use of simple, open-type furnaces, as very little reducing coal is added and virtually no gas is generated in the electric pig iron furnace. This means that furnace roofs, roof cooling systems, and gas cleaning systems, etc. are dispensed with and considerable investment costs saved.

As can be calculated from Figs. 15 and 16 (pages 57 and 58), the electric process is an economic proposition under certain conditions

when prereduced burden is used, even at power prices of 12.5 to
15 mils/kWh (0.05—0.06 DM/kWh) and coke prices of 15—20 $/t
(60—80 DM).

In other words, a possible reduction in the price of electric
power by building atomic power stations could establish new fields
of application for the electric low-shaft furnace process, especially
for the use of non-coking coal. Even where power costs are high,
the use of a prereduced burden and the possibility of using inferior
coals offer a promising future for this process.

It is stressed that iron and steelworks can combine high-capacity
blast furnace plant with smaller electric pig iron facilities. This
offers the possibility of using high-grade gas from the electric pig
iron furnaces (which would then have to be of the closed type) for
certain special purposes, and of using up fine reduction materials
too small for use in the blast furnace and which at the present
time are still difficult to process. In addition, certain proportions
of ore fines can be utilised (especially where open furnaces are used
and charged with prereduced material), and special grades of pig
iron can be produced without having to spend time and money
on converting a blast furnace for this purpose. Taken all round,
the electric pig iron process should be given very careful consideration in all countries that have no coking coal at their disposal, and
also in all cases where low-capacity plant is required.

c) Steelworks

General. Once the production rate and the production programme
of the steelworks have been determined, the buildings and layout
of the individual departments can be planned. The smooth supply
of pig iron, hot metal, scrap, coal, additions and auxiliary materials,
and the smooth outward flow of crude steel, slag, and skull is
extremely important, as is the free circulation of moulds. Supplies
of power, blast furnace gas, water, compressed air, steam and so
on usually present no special difficulties.

The sizes of the converters, open-hearth furnaces, electric furnaces, mixers, and ladles must be carefully matched. Any mistakes
made at this stage will make themselves felt during the entire life
of the plant, and adversely affect the performance of the entire
works. The planned ingot weights are taken as a basis; this includes the standard ingot and the heaviest ingot. For example,

for standard ingots of 4—5 t, a maximum ingot weight of 100 t, and an output of

> 600,000 t/p.a. basic Bessemer steel + blown metal
> +300,000 t/p.a. open-hearth steel
> +100,000 t/p.a. electric steel,

we could select

> 3 60-ton converters
> 3 100-ton open-hearth furnaces
> 2 40/60-ton electric furnaces.

The ladles must have a capacity of 60 t; these will also suffice for the open-hearth furnaces, which must then be tapped into two ladles at a time. The use of 100-ton ladles, which would necessitate heavier cranes and craneways, is justifiable only when 60/100-ton ingots are continually cast in large numbers. In the latter instance, the open-hearth steelworks would need a 150-ton crane.

1. Open-hearth Steelworks

Open-hearth Furnaces. The main or exclusive steelmaking process determines the type and size of the open-hearth furnaces, the output, and the steel grade.

If the charge is largely made up of hot metal or blown metal, ladle handling and furnace charging methods must be planned at the earliest stage. In bigger steelworks, hot metal will not be transported through the casting bay, and charging will wherever possible be effected from the platform side, keeping the distance to be covered by the crane to a minimum.

Tilting furnaces, which usually have capacities of some 100 t and more, are nowadays used in special cases only, e.g. where two slags are worked.

Actual furnace selection—i.e. more or less the selection of the furnace ends—is more a matter of personal taste. The "Maerz" furnace has in general the same output as furnaces with ends of the "Friedrich" type and similar. The "Moll" furnace, which has a refining effect, occupies a special and oft-debated position. The chambers of this furnace are of special design; consequently, the entire lower part of the furnace differs greatly from that of orthodox furnaces.

Generally speaking, the danger of slag being carried over from the hearth and blocking the regenerator checkerwork is not given sufficient attention; this danger is most pronounced in steelworks using solid charges and no heavy scrap. PAUL KÜHN, of Nieder-schelden, developed and tested a slag pocket many years ago that not only holds back slag entrained in the waste gases, but collects it in the molten state, a most important feature (Fig. 17). The slag collects in the hottest part of the pocket, and is thus kept liquid for easy removal. The limited number of columns built into the slag pocket to baffle the slag particles entrained in the waste gases are so sturdy that they last a full furnace campaign. This type of slag pocket has given good results.

The service life of the furnace roof is a major concern in open-hearth furnace operations. First-grade bricks, meticulous con-struction, careful flame control, and, above all, rigid control of

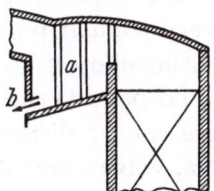

Fig. 17. OH Furnace end with preliminary
slag drain (designed by PAUL KÜHN).
a baffles; b liquid slag drain.

furnace operations to prevent overheating of the roof, can ensure a long service life. These factors are much more important than any "special" roof bricks; many have definite advantages to offer, but the factors listed above are decisive.

Open-hearth Furnace Fuels. Producer gas was used to fire open-hearth furnaces up to about 20 years ago, but is very seldom used for this purpose nowadays. It was made by gasifying hard coal, lignite briquettes, or crude lignite and supplied to the furnace in the hot state.

Many open-hearth furnaces are still fired with rich gases, i.e. coke-oven gas, natural gas and so on. However, 100% coke-oven gas is hardly ever used, as the flame is not luminous and heat transfer to the bath is poor. Heavy fuel oil is the main fuel in modern open-hearth shops; it is often mixed with coke-oven gas or natural gas in ratios of up to 50:50 (in terms of kcal).

The average fuel consumption rate per ton of crude steel (sound ingots) measured over the month, i.e. counting week-end hours and normal breaks in operation, lies between 0.9 and 1.4 million kcal/t in the case of larger furnaces operated on hot charges. In the case of smaller furnaces and when cold charges are used, the figure lies between 1.3 and 2 million kcal/t.

The addition of oxygen to increase output has proved successful in many cases, producing increases of up to 20% and fuel reductions of up to 25%. In some cases, four oxygen lances were introduced vertically through the corners of the roof, but this was not particularly successful as the service life of the roof was badly affected.

As opposed to this practice, the insertion of oxygen lances through the furnace ends produced good results. The oxygen volume used was about 30 Nm³/t.

Where producer gas is still used, possibly being mixed with rich gas in the furnace ends, the open-hearth furnace needs two gas chambers and two air chambers. The same applies when these furnaces are fired with a mixture of blast-furnace and coke-oven gas, a very popular practice at one time.

The firing of open-hearth furnaces with *cold coke-oven gas, natural gas*, and *oil* permits simplification of the furnace design, as the gas chambers can be dispensed with. The lower part of the furnace, the ducts, valves, and slide valves are much less complicated. Gas and oil are fed through the furnace ends together with carburising agents. At one time, the practice of firing open-hearth furnaces with cold coke-oven gas without carburising fuels was very popular in Germany, but has since been largely replaced by firing with coke-oven gas that has been substantially debenzolised. Cold coke-oven gas is therefore nearly always used with carburising fuels. Depending on the quality and analysis of the coke-oven gas, the carburising fuels must produce some 10 to 50 % of the total heat required.

In countries that have rich natural gas resources, such as the USA and Italy, and nowadays in the Netherlands and Germany, too, coke-oven gas is giving way to natural gas, which is cheaper and has a higher calorific value. Even though natural gas has a high methane content and produces a good luminous flame, it is still often mixed with fuel oil, in the same way as coke-oven gas. The highly luminous flame so produced ensures maximum heat transfer to the bath and, consequently, optimum fuel consumption rates.

The technical aspects of *oil-firing systems* need not be discussed in detail; the USA and Great Britain have been using such systems on an industrial basis for many years. The picture is somewhat different in the German iron and steel industry, where fuel oil did not achieve any importance until a few years ago. However, the use of oil has increased sharply in recent times. In 1952, the oil consumption figure was as low as 100,000 t; by 1964, this had shot up to some 2.5 million tons[1]. Of this, about 2/3 were used in open-hearth steelworks. Mineral oil contains some 3.5% of sulphur, an unwelcome "free gift" that must be treated with respect, particularly when making high-grade steels. The mineral oil industry is now also able to supply a fuel oil with sulphur contents of only 1.8% or less.

In integrated works having several different steelworks, e.g. an open-hearth works and a basic Bessemer works or an oxygen converter works, the majority of commercial steels will be produced in the latter; the open-hearth works will cater for special steel grades and ingot sizes. If the open-hearth process is used on an exclusive basis, the casting pit will cater for the bulk of the commercial steels and the special programme. The casting pit and bay must therefore be designed to cater for the various demands made, i.e. for a high output of standard ingots, heavy forging ingots, and a variety of ingot shapes and weights to be poured under specified conditions.

Steel-casting Foundry. Every iron and steelworks needs a certain amount of steel castings for repair work. Nevertheless, the expediency of establishing a steel-casting foundry next to the open-hearth works must be carefully examined. If an iron foundry is available anyway, the two can be combined. In any case, a steel-casting foundry needs a number of small electric furnaces with capacities of 5 to 30 t, or Bessemer converters with capacities of about 5 t; in larger works, it is expedient to provide one or two of the furnaces with acid linings.

General Remarks. In open-hearth steelworks, characteristic values for rough estimates are governed by the heating system, furnace type, furnace size, type of charge used, i.e. whether hot or cold, blown metal, and so on. However, values roughly corresponding to those given below may be expected at the present time; as shown in Fig. 18, the figures are based on actual results attainable with present-day furnace outputs.

Example An open-hearth steelworks is to produce 300,000 tons of crude steel per annum; of this, 50,000 tons are to be produced in small furnaces operated on

---

[1] See: Statistisches Bundesamt, Eisen- und Stahl-Statistik für 1964 (Iron and Steel Statistics for 1964, issued by the Federal German Statistics Office).

cold charges. 75,000 tons of hot metal p.a. can be expected. How should the steel-works be designed ?

Two 40-ton furnaces that can each produce 6 t/h or 40,000 t/year[1] are provided for the 50,000 t of high-grade steel. The remaining 250,000 t are to be produced in 140-ton furnaces operated on a 30% hot metal charge. The output of one furnace may be taken as being 19 t/h or 130,000 t/year, which means that two open-hearth furnaces are required.

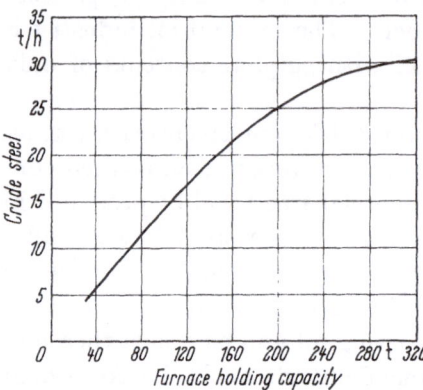

Fig. 18. Hourly ouputs of West European OH furnaces (values established in 1965) scrap-pig process (max. 350 kg pig iron/t).

Fuel consumption rates. These are to be calculated on the basis of the monthly average, and not on the usual basis of the operating hours at full furnace output. If for correct comparison the available heat at the furnace valve or the furnace end is always used, gasification must also be taken into consideration.

Depending on the work's capacity, the investment costs for an open-hearth steelworks amount to 25 to 37.5 $ (100.— to 150.— DM) per ton of annual capacity, the average figure being about 32.2 $ (130.— DM) (Basis: 1965).

2. Electric Steelworks

Electric steel is a quality product, the manufacture of which calls for a high degree of care and attention. Accordingly, the planning and building of electric steelworks also demands great care.

Arc furnaces and high-frequency furnaces are the mainstays of electric steel shops. Other types of furnace, such as single-phase, low-frequency and similar units are comparatively rare in works of this kind, and are therefore not discussed here.

Arc furnaces of the HÉROULT, RODENHAUSER and similar designs feature three carbon or graphite electrodes. The arc furnaces of the DEMAG-Elektrometallurgie (formerly Siemens & Halske) and Stein-Roubaix are the most widely used types in Europe. At the present time, these furnaces are built with capacities of up to 120/150 t for cold or hot charges. Various charging systems are

---

[1] This is based on 300 furnace working days or 7,000 furnace working hours p.a.

used by the different manufacturers, but no definite superiority can
be claimed for any particular design.

High-frequency furnaces are used for melting high-grade steels,
particularly tool steels, heat-resistant steels, acid-proof steels, and
other high-alloy steels. This type of furnace is built for capacities
of 100 kg to 8 t, the normal capacity being 500 to 2,000 kg. These
furnaces often permit the use of a common transformer and capa-
citor facility for two furnaces, so substantially reducing investment
costs and the space requirement.

Operation. High-alloy steels are usually made from a cold charge,
i.e. cold charges are always used with high-frequency furnaces.
Otherwise, hot charges are frequently used. These can take the
form of metal from the oxygen steel converter or the open-hearth
furnace, or blast-furnace metal can be used. In big electric
steel shops, all these charging possibilities—not forgetting 100%
cold charges—must be given due consideration; the cranes and other
handling equipment must be planned and designed accordingly.

Characteristic values are difficult to quote for electric furnaces,
as the various steel grades call for different charges and alloying
elements, which in turn produce varying melting periods and furnace
outputs. However, the following approximate values are given for
normal high-grade steels and electric arc furnaces:

Capacity [t]		up to 15	over 15	
Hot charge [%]		–	–	50 and more
Transformer rating	kVA/t	200–300	200–300	200–300
Furnace rating	t/h	2–5	4–6	6–10
Melting period comprising:	h	6–8	6–8	4–6
melt-down	h	4–5	4–5	2–3
refining	h	2–3	2–3	2–3
Power consumption comprising:	kWh/t	800–900	600–700	400
melt-down	kWh/t	500–600	400–500	200
refining	kWh/t	300	200–300	200

*Remarks on the Processing of Sponge Iron Produced to the Direct
Reduction Process to Steel in the Hearth Furnace*

As mentioned in the description of the processes (see page 33),
the gangue in the ore normally remains in the sponge iron during
both the gas reduction process and the rotary kiln process. During

the steel process, a slag ratio of $CaO:SiO_2 = 2.5$ is normal; the
liquid volumetric ratio of 1 t steel:1 t slag is 0.141 m³/t:0.4 m³/t,
or 1:2.8. This has various important consequences, as shown by
the relationship between the gangue in the sponge iron and the slag
volume (Fig. 19).

The centre curve shows that with some 63% Fe in the ore and
the sponge iron made from it, the steel to slag ratio is 1:1 when

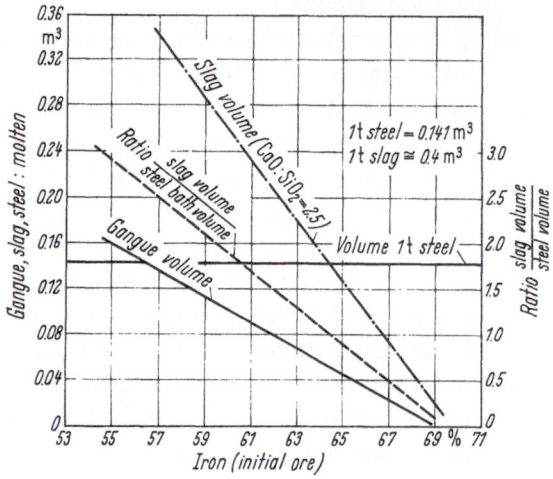

Fig. 19. Relationship between steel bath, gangue, and slag volumes ($CaO:SiO_2 = 2.5$)
when melting sponge iron, as a function of the Fe content of the initial ore.

this sponge iron is the only raw material used in the steel process.
If sponge iron made from an ore containing some 59% Fe is used,
the steel-to-slag volume ratio rises to 1:2.

In view of these facts, it must be stressed that great care should
be exercised when using sponge iron made from ores with less than
63% Fe, even when the amount used is cut to some 80% by using
20% home scrap.

The conditions are, of course, based on the fact that gangue is
normally acid; the conditions may be much more favourable in the
case of a self-fluxing gangue.

However, as numerous industrial-scale investigations and results
obtained in day-to-day operations using sponge iron have shown,
the use of sponge iron made from ores with Fe contents of less
than 63% produces certain changes; power consumption can rise

from the normal rate of 600 to 800 kWh/t and more, melting times increase, and the consumption rates for electrodes and refractory materials can rise substantially.

On the other hand, when using sponge iron with high percentages of Fe, it is very possible to achieve power consumption rates comparable to those usual when melting scrap. Of course, a special modus operandi is called for when using non-briquetted sponge iron, its specific weight being very low.

As far as sponge iron works that are to supply a steelworks on an exclusive or almost exclusive basis are concerned, it is most advisable to carry out suitable tests on an industrial scale in a steel furnace before a decision is taken. The favourable results obtained to date when using large amounts of sponge iron were in the main achieved in electric arc furnaces. It is, of course, also possible to charge sponge iron into an open-hearth furnace in loose form, at least in theory; the basically different form of heat transfer can produce difficulties, especially if the gangue has a relatively low melting point. These difficulties can be obviated to a high degree by briquetting the sponge iron.

3. Premelting Plant

The cupola furnace has attained great significance in the steel-works field, too, since the introduction of the hot blast, which is heated by the top gases. With the hot blast, steel scrap can be continuously melted down over long periods in basic-lined and unlined cupola furnaces without any difficulty whatsoever. The end product is in effect a synthetic pig iron with an extremely low content of manganese and silicon. Pig iron scrap, bloomery iron and so on can be transformed into a useful premelted iron. Experience has also shown that the use of such iron in open-hearth furnaces has its optimum at 50% (of the charge), at which value a 40 to 50% increase in furnace output is achieved. In comparison, the optimum charge percentage of pig iron in the open-hearth furnace when using the scrap — pig iron process is 30/40%; again, the use of hot metal instead of pig iron produces an output increase of approx. 20% only.

In turn, this increase in output substantially reduces the furnace operating costs, as these are governed to some 75% by the time factor. A further advantage is the actual increase in production

5*

and therefore turnover, which in many cases is of decisive importance[1] (Fig. 20).

Premelting furnaces are, of course, mainly intended for open-
hearth steelworks which must use a 100% cold charge. However,
cupola furnaces that produce synthetic pig iron can also be used
in conjunction with oxygen steelworks[2].

As premelting plants are relatively inexpensive to build, they
can effect a considerable reduction in the overall investment costs
required when extending or building open-hearth steelworks. The
following calculation is given as an example:

	Unit	Normal OH steelworks (100% cold charge)	OH steelworks with premelting plant
Steelworks capacity	t/year	100,000	140,000
Specific furnace output	%	100	140
Investment costs			
OH steelworks	$	3,500,000	3,500,000
	DM	14,000,000	14,000,000
Premelting plant (70,000 t/p.a. ×9 $/t (36 DM/t)	$	—	600,000
	DM	—	2,400,000
Totals:	$	3,500,000	4,100,000
	DM	14,000,000	16,400,000
or	$/t	35.0	29.30
	DM/t	140.0	117.0
or	%	100	84

This capacity increase of 40,000 t/y costs 600,000 $ (2.4 million
DM), or 15 $ (60.— DM) per ton.

In the same way as the open-hearth steelworks, the premelting
plant can also work in conjunction with oxygen converters and
electric arc furnaces. In the latter case, the electric furnace would
have to be operated in accordance with the "von Roll" process
(pig-iron—ore process) or with oxygen.

---

[1] VOIGT, H.: Der basische Heißwind-Kupolofen im gemischten Hüttenwerk.
(The Basic Hot-blast Cupola Furnace in Integrated Iron and Steelworks). Stahl
und Eisen 78 (1958) pp. 284/91.

[2] RICHTER, A., COHNEN and P. JAKOBI: Der basische Heißwind-Kupolofen,
das Siemens-Martin-Stahlwerk und der Oxygen-Tiegel als Verbundeinrichtung.
(The Basic Hot-blast Cupola Furnace, the Open Hearth Steelworks, and the Oxygen
Steelworks as an Integrated Plant). Stahl und Eisen 78 (1958) pp. 273/84.

4. Blown Steel Processes

aa) Basic Bessemer Steel Plant

Converters. The basic Bessemer process is used in conjunction with hot metal containing more than 1.8% P. Converter sizes are established during the planning stage, taking the open-hearth steelworks and the electric steelworks into due consideration. In Europe, most converters are still cylindrical in shape; H. A. BRASSERT

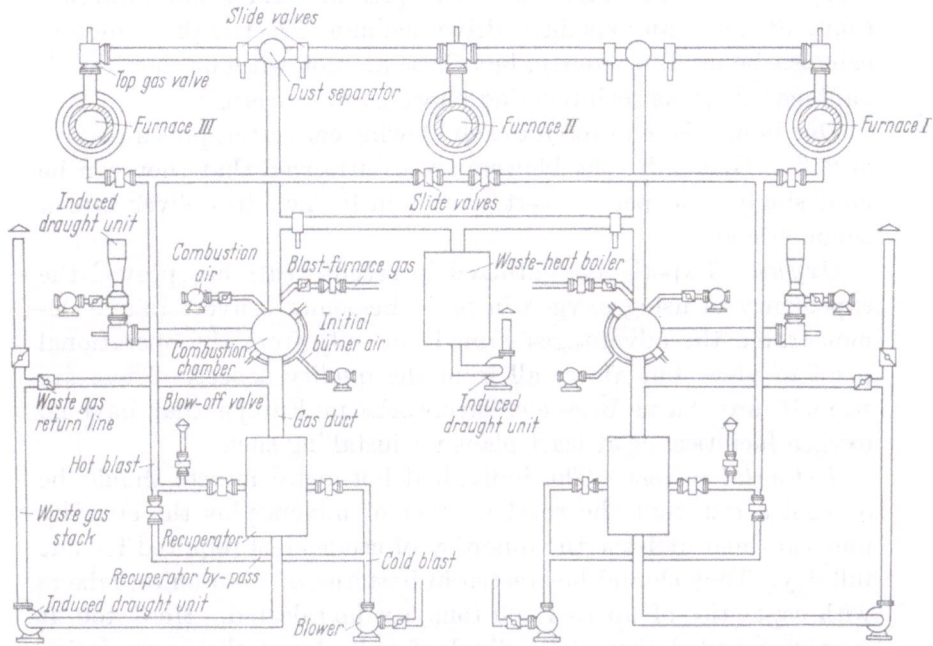

Fig. 20. Hot-blast cupola premelting plant for the OH steelworks of the Salzgitter iron and steelworks. 3 cupola furnaces, each with a capacity of 25—30 t/h.

introduced the round converter in the Reichswerke, Salzgitter, but the expected advantages were realised to a limited extent only. The increased cross-section, the resulting deeper bath, and the greatly increased space above the bath did in fact increase the output, and also permitted the blowing of hot metal with Si contents in excess of 1% without increasing spitting. Round converters, which have a cross-section equal to that of an orthodox converter after $\frac{1}{3}$ to $\frac{1}{2}$ of its campaign, offer the advantages of a run-in converter from the first charge on. However, the lining did not give satisfactory service; the result was a return to the orthodox cylindrical lining, in spite of the other advantages offered.

Basic Bessemer steel plants should be equipped throughout with converters of the same type and size wherever possible. Exceptions should be made only when really pressing considerations make this necessary.

Blowers. The converters can be supplied with air by turbo or reciprocating blowers. Turbo blowers are usually given preference today, as they can better cope with pressure and volume fluctuations. Steam is an expedient drive medium. Electric drive motors can also be used, of course, but these are not without their disadvantages as far as reciprocating blowers are concerned.

One blower is required for each blowing converter, plus a reserve of 25%. Generally, the blowers are so arranged that they can be used singly, one per converter, or jointly, i.e., to deliver into a common mains.

Oxygen. Experience[1] gathered in recent years has proved the expediency of using oxygen in basic Bessemer converters and demonstrated the advantages gained not only from the operational point of view, but above all from the quality aspect. There are scarcely any basic Bessemer steelworks in Europe that have no oxygen facilities, or at least plans for installing such.

Hot-metal mixers. The individual hot-metal mixers should be of such a size that the total number of mixers plus the standby unit can hold at least the quantity of crude steel required for one full day. They should be erected in batteries of 2—3 units; mixers with capacities of up to 1,200 tons can be selected. Here, too, it is recommended that the individual units be of the same design and capacity.

If blast-furnace gas is the sole heating medium, the air *and* the gas have to be preheated; it is therefore much better to use mixed gas or coke-oven gas with added oil or coal dust.

---

[1] Cf. W. BADING: Stand und Entwicklung des Windfrischverfahrens. (The Air Refining Process — Present State and Development.) Stahl und Eisen 71 (1951) pp. 373/88. — MEYER, K., H. KNÜPPEL, H. POTTGIESSER: Versuche zur Erschmelzung stickstoff- und phosphorarmer Thomasstähle. (Trials on the Manufacture of Basic Bessemer Steels with Low Nitrogen and Phosphorus Contents.) Stahl und Eisen 72 (1952) pp. 225/32. — DICK, W.: Die Möglichkeit der Verbesserung des Thomasstahls durch Sauerstoffzusatz zum Gebläsewind. (The Possibility of Improving Basic Bessemer Steel by Adding Oxygen to the Blower Air.) Stahl und Eisen 72 (1952) pp. 233/42.

Tilting mixers are now the standard type, of course, and flat-hearth mixers are used in special cases only.

Mobile mixers of the torpedo type are used to transport hot metal from the pig-iron plant to the steelworks; some large-capacity torpedo mixers are nowadays using public railways.

Desulphurisation Plant. During the past 30 years, more and more "acid" ores have been smelted to the Paschke-Peetz process, and the basic Bessemer pig iron so manufactured must be desulphurised. Earlier, soda or a mixture of lime, fluorspar and soda was used for desulphurisation in the blast furnace itself, but it was later established that in many cases it is more expedient to carry out desulphurisation in the steelworks.

Experience gathered in the Salzgitter iron and steelworks of the Reichswerke shows that the sulphur content of blast-furnace metal can be substantially reduced by pouring the hot metal into the desulphurisation ladle from heights of 10 metres and more. For this operation, a crane must be mounted on a stationary structure to raise the ladles coming from the blast furnace. The ladle is then emptied into a desulphurisation ladle positioned under the crane hook.

Basic Slag Plant. Basic slag plays an important part in the economic structure of the steelworks. The slag is crushed and sacked for dispatch; a certain degree of intermediate storage is necessary to cater for the longish interruptions in dispatching caused by the seasonal demand and transport problems (waterways). As it is not advisable to store crushed basic slag in sacks for long periods, an intermediate store yard should be built for lump or granulated slag. This buffer store should be big enough to take the slag accumulating over 2—3 months.

Facilities have been built for the dry granulation of liquid slag, but the results obtained to date have not been encouraging. Nevertheless, this question must be followed up if the slag crushing installations are to be relieved.

No substantial improvements have been effected in crushing plant for basic slag during the past few years, and new facilities will consequently follow the established pattern.

The only really new features are the intermediate store yard for lump slag and the capacity increase in the crushing plant and all following machinery up to the loading facilities to cater for the peak load when the seasonal demand is high.

Dolomite Plant (for converter bottoms). This important auxiliary shop should be located as close to the converters as possible to cut the circulation times for converter bottoms and converter relining times to a minimum. It should be of such a size that it can never constitute a bottleneck.

Casting Bay. Basic Bessemer steelworks are usually designed for the production of large amounts of steel of one grade in definite ingot sizes, which means that in a lot of cases it will be possible to employ the top-pouring method, the moulds standing on wagons, the casting ladle remaining in the one position. This method keeps the space requirement down, and permits the mould fettling shop to be erected in a different area; where the blooming mill and soaking pits are located a considerable distance away from the steelworks, the moulds can also be stripped near the soaking pits. When this method is used, careful attention must be paid to the smooth circulation of the wagons, ingots, and moulds. If this is ensured, however, the efficiency of the basic Bessemer steelworks will no longer be impaired by the casting bay.

Mould Shop. Wherever possible, the mould fettling shop should always be located in a separate bay parallel to the mould circulation path. The advantages of smooth and unimpeded mould fettling in a separate bay are self-apparent and need no further discussion.

Output. The annual output per ton of converter capacity can be estimated at 5,000 to 7,000 tons of basic Bessemer steel; in isolated instances, annual outputs of up to 9,000 tons of steel per ton of converter capacity may be achieved.

Accordingly, a basic Bessemer steel plant with $4 \times 60 = 240$ t converter capacity produces some 1,200,000 to 1,680,000 tons of crude steel per annum. The investment costs can be taken at some 25 \$ (100 DM) per annual ton (Basis: 1965).

bb) The Oxygen Process. This process was developed in the Linz iron and steelworks with the cooperation of K. DURRER[1] for the blowing of phosphorus-free steelmaking pig iron with pure oxygen. The process is based on the blowing of pure oxygen (min. 99% O_2) at high speed (up to 100 m/s) onto the surface of a hot-metal bath in a converter. This process produces a good steel which is in many

---

[1] Cf. DURRER, HELLBRÜGGE, and RICHTER-BROHM: Die Entstehung des "LD" Sauerstoffaufblas-Verfahrens (The Development of the "LD" Oxygen Top-blowing Process) Stahl und Eisen 85 (1965) pp. 1751/54..

ways the equal of open-hearth steel. The oxygen process is used all over the world, and the total world capacity has now reached 60 million tons per annum (1965).

It is now also possible to use the oxygen process in countries that have always depended on the basic Bessemer process for their phosphorus-bearing ores. In the special processes developed in Germany and France for this purpose, the LD-AC and OLP processes, fine limestone powder is entrained in the stream of oxygen and blown into the bath. In special cases, finely-ground lime is added to the bath separately. Whereas the oxygen process can be used for P contents of up to 0.5% without any difficulty whatsoever, P contents of between about 0.5 and 0.8 call for the normal top-blowing process and two slags, and P contents in excess of 0.8% can only be dealt with by the LD-AC process. The working of two slags and the introduction of lime, which needs careful supervision, naturally lengthen the tap-to-tap time and therefore reduce output in comparison to plants of equal size.

The oxygen steelworks should normally be equipped with at least two converters, one always in operation while the second is being relined. If the furnaces and similar equipment are kept in operation over weekends and holidays, a normal practice in many countries, the plant can be worked for some 8,000 hours per annum. On the oxygen steelworks being expanded, 3 converters would be the optimum figure, as the third converter would exactly double the capacity of the 2-converter shop. In theory, there is naturally nothing to prevent an increase to 4 or 5 converters. However, experience gathered during the past few years shows that the simultaneous operation of three or more converters has adverse effects on the supply of hot metal, the charging of scrap and additions, and converter tapping. The result is an unavoidable drop in output. Due to this fact, no oxygen steelworks for up to 5 million tons p.a. erected in recent years has more than 3 converters; the biggest of these, (1965) with capacities of 300 tons, are to be found in the USA and in Taranto. As higher production rates are needed there, and as the firms concerned do not wish to increase the converter capacities beyond certain limits, two separate oxygen steelworks, each with 2—3 converters, are to be built.

With the normal basic Bessemer process, some 50 to 100 kg of cooling scrap can be charged per ton of hot metal, but with the oxygen process this figure increases to 100—250 kg.

Again, a certain proportion of the scrap can be replaced by pellets with high Fe contents. In other words, the oxygen steelworks is much more flexible as regards the use of scrap and in the ratio of scrap to hot metal.

Generally speaking, the refractory lining of an oxygen converter lasts for some 300 blows, but some steelworks are known that continually achieve 500 blows.

The average tap-to-tap time for the oxygen converter, as measured over a period of twelve months, (i.e. including unavoidable breakdowns etc.), is 40—60 minutes. The tap-to-tap time for small oxygen converters is rather less than that of bigger units. Again, the number of converters used affects the output figure; for example, steelworks using two converters, of which one is always in operation, have higher outputs than shops with three converters, two of which are always in operation. This output drop can be expressed as an operating ratio, or simultaneity factor:

Converter capacity t	Simultaneity factor
20	0.95
90	0.94
160	0.93
230	0.92
300	0.91

Table 5. *Oxygen Steelworks*

Converter capacity	Charges per day	Tap-to-tap time (min.)	Charges per day	Tap-to-tap time (min.)
	One converter in operation		Two converters in operation	
30	34.7	41.5	32.9	43.5
50	33.9	42.5	32.0	45
100	32.2	45	30.2	48
150	30.3	47.5	28.3	51
200	28.6	50.5	26.4	54.5
250	26.7	54	24.5	59
300	25.0	57.5	22.7	63

The factors for intermediate converter capacities can be interpolated.

The simultaneity factor means that the output of two converters in actual operation is lower than that of one (operating) converter of the same size.

Example   1 160-ton operating converter produces 1,440,000 tons per annum
          2 160-ton operating converters produce 1,440,000 × 2 × 0.93
            = 2,678,000 tons per annum

Fig. 21 gives the capacities for one operating converter and two operating converters (the latter with the dotted line), based on these considerations and values, as a function of the actual number of operating hours per annum. For planning work, either the operating days p.a. (7,200 hours per annum) can be used for calculation purposes, or an estimated tap-to-tap time of 60 minutes and a higher number of operating hours (e.g. max. 8,000 hours p.a.) can be taken for converters with capacities of 200 tons and more. As

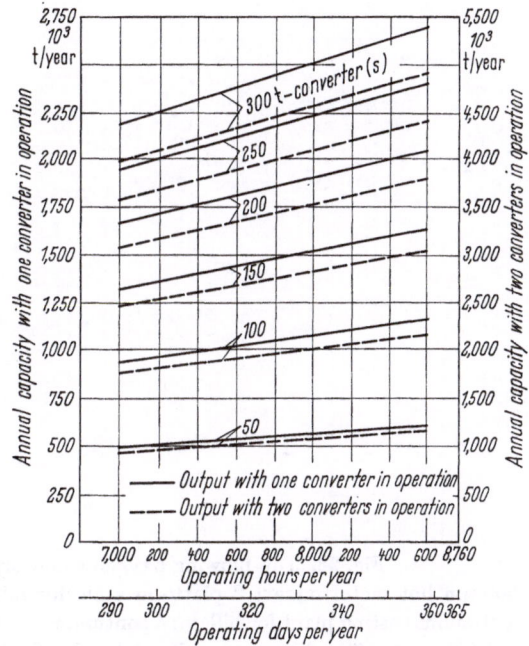

Fig. 21. Annual steel capacity of one and two oxygen-blown converters (BOF) as a function of the operating time.

the operating crew becomes more skilful, it may prove possible to cut the tap-to-tap time by 10%, for example, which would increase the annual output by the same figure. A further reserve is available in that the converters can hold more hot metal as time goes on (lining wear), which also increases output.

The non-tilting converter is tapped in rather the same way as an OH furnace; this design obviates the traditional and expensive tilting mechanisms, for example.

The steelworks oxygen plant must be designed to produce some 60 to 70 Nm^3 O_2 to cover the demand of about 50 to 60 Nm^3 O_2/t. In industrial areas with several neighbouring iron and steelworks, it often proved advantageous to build one central oxygen plant, which can also market oxygen in bottles. This type of plant is a more economic proposition than a small facility for one or two converters; in the latter case, the demand for oxygen is not constant, whereas the larger plant can use storage tanks.

As a rule of thumb, it may be taken that one Nm^3 of oxygen costs the same as 2 kWh.

It need hardly be emphasised that the oxygen converter sizes should be matched with the remaining equipment, i.e. ladle capa-

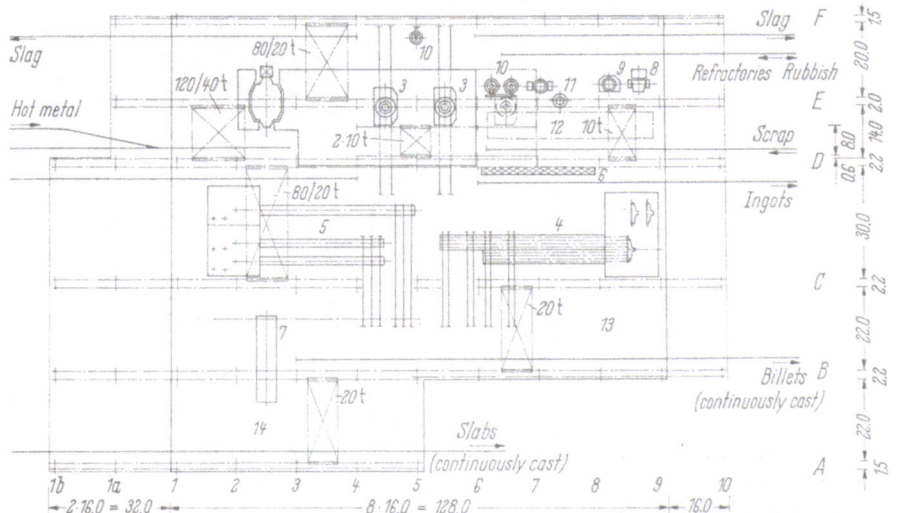

Fig. 22. Top-blowing oxygen steelworks. Plan view.
1 600-ton hot metal mixer; *2* platform scale for mixer; *3* 50-ton LD converter; *4* continuous casting plant for billets; *5* continuous casting plant for slabs; *6* teeming stand for ingots; *7* cooling section for slabs; *8* ladle lining breaking stand; *9* ladle lining pit; *10* ladle heating facility; *11* stopper rod furnace and changing stand; *12* scrap area; *13* billet area; *14* slab area.

cities, crane capacities, and hot-metal mixer capacities. If 100-ton ladles are used, it would be advisable to use 100-ton converters; converters with capacities of 200 tons could also be used, but it would be most advisable to avoid intermediate sizes.

A very important factor for all converting steel mills is the legislation on air pollution control passed in many industrialised countries in recent years. These laws required that dust collecting equipment be used for all types of blowing converters to remove all but the last trace of dust from the converter gases. The existing

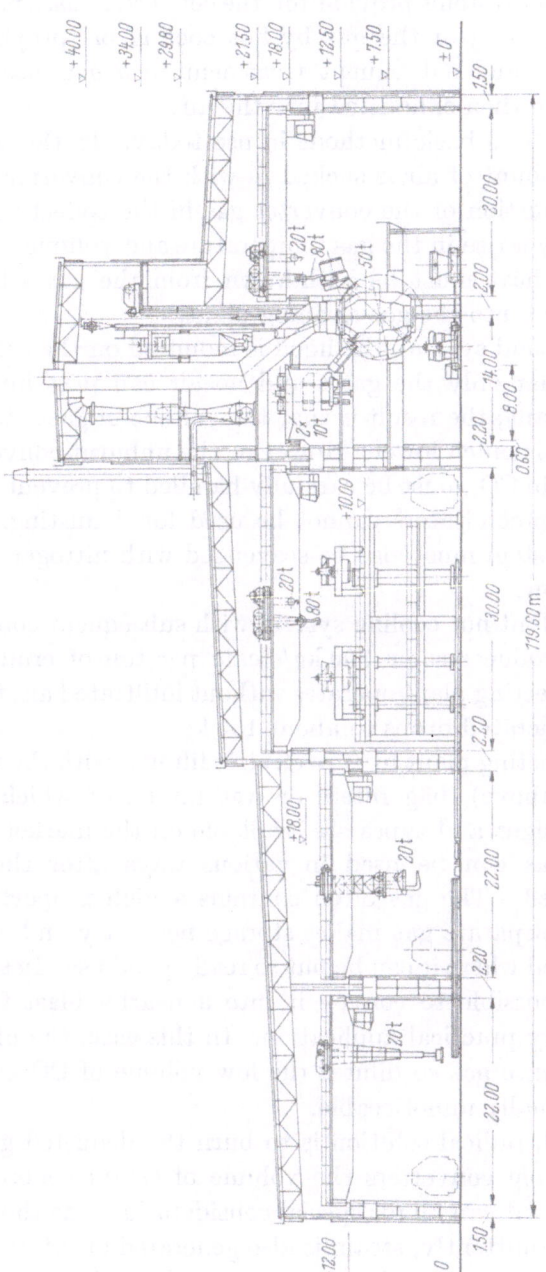

Fig. 23. Top-blowing oxygen steelworks. Sectional view.

dust collecting systems provide for the converter gases being caught in a hood, cooling of the gas by hot cooling or spraying to some 400—500 °C, and subsequent treatment in a gas cleaning plant. The gases are then exhausted into the air.

There are two basic methods in use today. In the one system, a certain amount of air is sucked in with the converter gases; this causes combustion of the converter gas in the collecting hood and consequently a rise in the gas temperature and volume. As a result, much more heat must be withdrawn from the gases before they can be passed into the gas cleaning plant.

In the second system, the hood is mounted on the converter top to ensure that only the gases and no air can pass into the dust collecting plant; the result is that the volume of gases to be cooled and cleaned is much lower. However, the unburnt converter gases, which contain CO, must be carefully handled to prevent explosions, i.e. electric precipitators cannot be used for dedusting. The dust collecting system must also be scavenged with nitrogen before and after the blow.

The La Mont hot cooling system with subsequent combustion of the gases produces some 300 kg/steam per ton of crude steel. If the dust collecting plant operates without infiltrated air, the amount of steam generated drops to about 150 kg.

Dust collecting plant feature electric filters, (with the reservation mentioned above), bag filters or wet filters, of which numerous different designs and types are available on the market. Unburnt converter gas can be used in various ways after the dust has been removed. The gas often contains a high proportion of CO; its use as a separate gas makes storage necessary and can only be recommended where it can be put to really good use. In some plants it may be possible to convey it into a nearby blast furnace gas mains, a very practical application. In this case, the high volume of blast-furnace gas so dilutes the low volume of CO gas that the latter is virtually unnoticeable.

The third, radical solution is to burn the dedusted gas. In the case of the big converters the volume of steam generated in the dedusting plant can all for special consideration. As the converters operate intermittently, steam is also generated at intervals, so that in spite of the pressure of up to 50 atmospheres it can only be used after being stored in a Ruths accumulator. This gives a uniform feed of steam, but cuts the pressure to a maximum of 10 atmospheres.

cc) Kaldo Process The Kaldo process, developed by KALLING, is also an oxygen refining process. It uses an inclined converter that rotates at about 100 rpm.

Compared with the oxygen (LD or BOF) process, the Kaldo process permits the addition of much more scrap per ton of hot metal (up to 400 kg/t), and limestone can be added instead of lime; removal of the carbon dioxide improves the refining action. The advantage of the Kaldo process lies in the fact that it has the highest metallurgical efficiency of all steel processes, i.e. the best grade of steel can be produced to this process. However, not only are the investment costs some 50% higher than those of oxygen (LD) plant, but the wear rate is higher, i.e. the lining service life equals only 30 to 50% of the LD converter lining life. In turn, this reduces the output to about 50—60% of that of an equally large LD plant.

d) Finishing Process

1. Continuous Casting

The idea of continuously casting steel in an endless strand is a century old, and some 40 years ago, aluminium and other non-ferrous alloys were cast in this way. During the past 25 years or so, tremendous advances have been made in the continuous casting of steel, and the process can now be regarded as being commercially workable.

Modern continuous casting machines can produce billets ranging from 70×70 to 300×300 mm, and slabs up to $140 \times 1,200$ mm, and even now, some facilities can produce slabs measuring up to $220 \times 2,000$ mm, and there is every sign that bigger slab cross-sections will be possible in the future.

Some very promising investigations have been carried out into the continuous casting of hollow bodies for tube manufacture.

Originally, continuous casting machines were of the vertical type and measured up to 30 m in height; in Russia, these machines were installed in pits up to 17 m deep to reduce the height of the tower above the shop floor. The latest development is aimed at reducing the overall height as such, and in Western Europe various solutions have been arrived at. One alternative is the vertical plant with a bending device to take the solidified strand from the vertical to the horizontal plane; other alternatives are the bow-type or oval casting machines. The vertical plant with strand bending facility reduced the overall height of the tower by about $\frac{1}{3}$, i.e. by some 10 m, and bow-type and oval casting machines by $\frac{2}{3}$ and more, i.e. the overall height of the latter machines lies between about 6—10 m. Bending of the strand must be effected when the strand is still so hot that the surface has solidified but the core is still pasty, if not in some measure liquid. Experience gathered to date indicates that these machines are perfect for smaller cross-sections. However, machines for bigger cross-sections, and particularly big slabs, are still in the pilot stage. In contrast to Western Europe, the USA and the USSR still use vertical machines for big cross-sections and slabs.

Rimming steel still poses some problems in the continuous casting field, and various possibilities of overcoming the difficulties are being investigated, e.g. by using degassing equipment ahead of the continuous casting plant. This steel is still in demand for certain applications, e.g. various types of plate and wire; in Germany, for example, some 25% of all rolled products are made from rimming steel. This demand cannot yet be catered for by continuous casting machines. Efforts are also being made to establish whether and to what extent rimming steels low in silicon can be replaced by steels killed with aluminium.

In order to guarantee continuous operation, continuous casting machines must have a minimum of two strands, one of which is made ready for the next cast while the first is in operation. This takes about 20—30 minutes.

Continuous casting operations are also influenced by the supply of liquid steel. The casting time is limited to a maximum of about 60 minutes; beyond this point the steel would begin to freeze in the ladle. Full utilisation of a continuous casting plant is relatively easy when using hearth furnaces, e.g. open-hearth or electric arc furnaces, provided the operations are well coordinated; the situation is much more difficult when the steel is supplied from an oxygen steelworks, the tap-to-tap time of the converters being 40 to 60 minutes. In these cases, two continuous casting machines must be used per converter to ensure smooth operations. The number of strands per machine is also governed by the relationship between the strand dimensions and casting rates and the ladle capacity and casting time (see above). As the steel casting temperature is set within very narrow limits, basic Bessemer steel cannot yet be continuously cast, having very high temperature fluctuations.

The refractory linings of the casting ladles and tundishes must be given special attention, as the wear rates are particularly high here.

Nowadays, stopper ladles are normally used for continuous casting operations, the heat losses being lower than those of tilting ladles.

The continuous casting machine yield is about 5 to 10% higher when compared with a product of the same size rolled down from an ingot in a blooming or slabbing mill.

Whereas ingots often have to be rolled out to 10—15 times their initial length—for rails, a 20-fold elongation is specified—experience indicates that continuously-cast material need not be rolled out to this degree in many cases.

Continuous casting plants can be of combined design, i.e. for both square sections and slabs.

The power requirement of continuous casting plants varies between 5 and 20 kWh/t, depending on the design; cooling water consumption equals about 3 m³/t, and approximately 3 Nm³ of oxygen are needed per ton of steel for the cutting torches.

Casting rates are difficult to specify, as they are in great measure dependent on the size of the cross-section being cast; for example, a strand measuring 300×300 mm is cast at the rate of 4 m/min, or 18 t/h.

The investment costs also vary with the plant capacity, strand dimensions, and plant design, i.e. whether vertical or bow-type. In general, larger plants can be estimated as costing approximately 15 to 20 $ (60—80 DM) per ton capacity. A comparison of prices shows that those of a continuous casting plant and a slabbing mill are even at a capacity of about 1.5 million tons. In other words, the rolling mill becomes much less expensive as capacity increases, whereas the continuous casting plant investment costs remain practically the same as further strands are added on the unit construction principle.

2. Rolling Mills

Rolling mill equipment is the most highly mechanised and costly part of the entire iron and steelworks, and must therefore be given very close attention during the planning stage.

As errors in this sector would have very far-reaching effects, the description of the various types of rolling mills available is preceded by some general information on rolling mill plant.

aa) Rolling Programme. The basic aim of the planning work must be to ensure a definite degree of utilisation of the actual rolling mill capacity. This is part of the reason why big new iron and steel-works install modern, fully-automatic, single-purpose rolling mills with capacities of several millions of tons p.a. (e.g. fully continuous wide strip mills). Of course, the market demand must be such that a high degree of mill utilisation is guaranteed at all times, as the service of capital for plant of this type is very high and the production costs would be adversely affected if the mill where to be under-employed.

Smaller works, particularly those in developing countries, have a critical problem to contend with. Normally, import statistics are used as a guide when planning iron and steelworks, the object being the domestic production of items previously imported, allow-ance also being made for future demands. However, these statistics nearly always cover a very wide range of rolled products, and full coverage would call for the most varied types of rolling mills.

If these were to be built, they could not—at least for the greater part—be fully utilised, the specific share in the market of the indi-vidual products not being big enough; also, for technical reasons, certain types of rolling mills, e.g. rail and heavy section mills, have a *minimum* capacity that could by no means be fully exploited.

As an example, we use here a planning requirement of approxi-mately 50,000 t/y of rails and heavy sections (low proportion of rails), 20,000 t/y of medium sections and 20,000 t/y of light sec-tions, reinforcing bar, and wire rod. Full coverage of this demand would call for a heavy section mill utilised to about 30%, a medium section mill utilised to about 40%, and a light section mill of simple design (for the low production rate) with very high processing costs (high labour costs); this also takes into account the rolling of the initial sections for the medium-section and light-section mills in the heavy mill (no continuous casting plant).

In this particular instance, a medium-section mill of heavy design would cover about 70% of the requirement, and, being fully utilised, would earn a profit.

In cases such as this, full utilisation of the plant capacity must be the keynote of all planning work; it must be accepted that the mill will produce a bigger tonnage of certain products than the country needs at the moment, and that other products will still have to be imported. Full utilisation of the plant keeps the pro-duction costs down; the surplus tonnage can be sold to secure foreign exchange for the purchase of items still in the import list.

This situation must also be borne in mind when deciding between low-capacity rolling mill plant with a lower profit-earning capacity and modern, high-performance mills with lower production costs.

The question as to which steel grades should be included in the production programme also merits close attention. Without going into detail, mention is made of the fact that high-grade and alloy steels call for much more extensive speed control systems (especially for lower speeds) and roll-pass designing than do normal commercial grades.

In addition, these steel grades call for extensive intermediate cooling, annealing, and processing facilities that not only incur heavy investment costs, but also increase the space requirement; the latter point must be given due consideration during the planning stage.

All things considered, rolling mills designed for commercial steel grades, particularly high-speed high-performance mills with fixed stock reductions (continuous blocks driven through gearings), are hardly ever suitable for rolling high-grade and alloy steels.

By the same token, mills designed specifically for high-grade and alloy steels are usually not suitable for the economic rolling of normal commercial-grade steels.

The future proportion of high-grade steels to be rolled in an ordinary mill must be examined very closely; it may be necessary to delete certain grades from the programme altogether, or the decision may be taken to build a more expensive high-grade and alloy steel mill and reach full utilisation of this plant by rolling a certain percentage of normal grades at increased production costs.

bb) Degree of Mechanisation. It follows from the statements made above that the degree to which a rolling mill is mechanised must be the subject of careful thought.

Two examples are given:

Example 1. A non-mechanised light-section mill with an annual capacity of 12,000 tons costs approximately 75 $ (300 DM) per ton capacity fob; a semi-mechanised mill with a capacity of 100,000 tons p.a. costs approximately 59 $ (236 DM) per ton capacity fob; a fully-mechanised rolling mill with a capacity of 250,000 tons p.a. costs about 40 $ (160 DM) per ton capacity fob. This example clearly demonstrates the advantage of highly-mechanised rolling mills. The absolute investment costs fob are 1:6:10 at a capacity ratio of 1:8.5:21. In addition, mechanisation and automation definitely result in higher profitability where labour costs are high.

6*

Fig. 24 shows the influence full and part mechanisation have on the production costs per ton for light-section rolling mills. These are average values taken from a selection of light-section mills built in 1961—65. The list is taken from Dr. Hubert Müller's well-known work on the subject[1].

The chart shows that the production costs lie around 12.50 $/t (50 DM) in the case of partly-mechanised mills with rather low outputs of between 15 and 30 tons per hour. The high hourly operating costs of mechanised rolling mills permit processing costs of the same order (12.50 $/t) at high hourly production rates only, and these costs can increase by up to 100% as the hourly production rate drops.

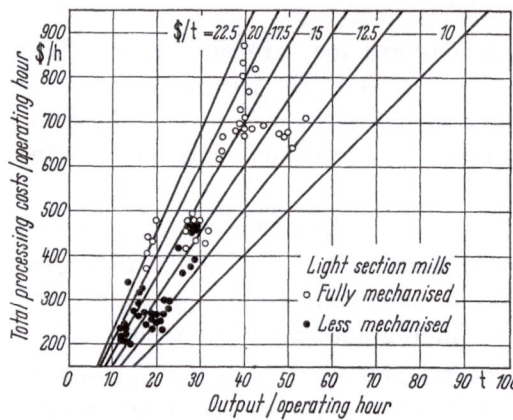

Fig. 24. Total processing costs in $/h and $/t as a function of the output per hour (t/h) for light section mills (average values; basis 1961—1965).

Example 2. A strip mill with one 3-high and two 2-high stands has a capacity of approx. 50,000 tons of strip p.a., using sheet bars (average strip gauge 0.8 mm); a modern, high-performance, fully-automatic hot wide strip mill has forty times this annual capacity. The fob costs of the lower capacity mill, including the facilities for rolling sheet bars from continuously-cast slabs, amount to approx. 165 $ (660 DM) per annual ton capacity as opposed to about 25 $ (100 DM) per annual ton capacity for the high-performance wide strip mill (less finishing line equipment and subsequent processing). In this case, the absolute fob investment costs are about 1:3 at a production ratio of 1:40.

The object of the examples outlined above is to draw attention to the necessity of carefully coordinating the rolling programme, investment costs, and degree of mechanisation.

cc) Starting Material for the Finishing Mills. Ingots and slab ingots can be taken from the steelworks, reheated, and put straight through the rolling mill. On the other hand, blooming mills, slabbing mills, or continuous casting plant can be used to provide the rolling mills with pre-rolled or continuously-cast material (e.g. blooms, billets, slabs).

---

[1] MÜLLER, H.: Wirtschaftlichkeitsrechnungen auf Hüttenwerken. (Profitability Calculations in Iron and Steelworks.) Stahl u. Eisen 85 (1965) p. 4, Fig. 3.

As far as the investment costs are concerned, the most inexpensive procedure to adopt is to transfer small ingots from the steelworks to the mill via pusher-type furnaces, the bigger ingots going to the mill via the soaking pits to conserve the heat still in the material.

Although the erection and operation of blooming mills, slabbing mills, and continuous casting plant entails additional investment and production costs, the following advantages are gained:

1. The use of pre-rolled or continuously-cast material cuts the scrap rate (croppings in particular) by 8—15%, which in turn increases mill output; mill stoppages are fewer (no gassy ingots, for example).

2. The end products have a better surface quality.

3. Rectangular initial material (as opposed to tapered ingots) helps to increase output.

Some important considerations are given below in regard to the question of using cast ingots or pre-rolled ingots (blooms).

As the stock runs rather slowly through low-capacity rolling mills (certain degree of manual operation), breakdowns can be kept within reasonable limits, whereas fully-automatic continuous mills demand perfect, pre-rolled initial material. From the point of view of service on capital, the use of a roughing mill or a continuous casting plant will scarcely be profitable. This is especially true when rolling reinforcing bars, for example, on which no great demands are made as far as surface quality and dimensional accuracy are concerned, as high yields can be achieved when rolling from an ingot.

Where light-section mills with capacities of between 30,000 and 50,000 tons per month are concerned, the use of continued casting plant is justifiable.

More complicated mills, e.g. those turning out a certain proportion of heavy sections and rails, can also be used to produce billets for the light-section mill.

One new trend is to use continuously-cast blooms instead of even heavy ingots, which have no quality limitations as far as rolling in heavy mills is concerned, in order to increase production, cut the scrap rate, and save the passes normally required to remove the taper of cast ingots. The saving of these passes—usually 2—is the new argument in favour of this procedure.

High-performance forging hammers arranged one behind the other (tandem) can be used to forge 50×50 mm billets from ingots to feed low-capacity rolling mills, or to meet a demand for small starting section in cases where the use of roughing mills or continuous casting plant would not be profitable.

The basic types of rolling mills are summarised below.

dd) Semi-finished Products Mills. If continuous casting plant is not used and the rolling mill plant in question has a reasonably high capacity, the ingots produced in the steelworks must be bloomed, or reduced to sheet bar, as the case may be. Slabs for plate and sheet mills of higher capacity are dealt with in the section on plate and sheet mills.

A fully-continuous billet mill should be erected in all cases where a steady demand for a high tonnage of billets exists.

Where the demand is less than about 250,000 tons per year, this material should be rolled in a two or three-stand heavy section mill; if required, a blooming stand can be erected ahead of the section mill.

This type of mill can generally also be used for rolling billets and sheet bars, as well as heavy finished products.

The final billet dimensions must be given close attention; the rolling of 50×50 mm billets, for example, can substantially limit the mill capacity, especially in the case of open mills, and heavy ingots would produce very long billets in this dimension range (no problem in continuous mills).

On the other hand, the use of heavy starting billets in light section mills calls for additional continuous stands; in the case of open mills, independent roughing stands will be needed in many cases.

Detailed descriptions of the various types of rolling mills would greatly exceed the scope of this book, and reference is made to the rich selection of special publications available on this subject. The information given below is, therefore, restricted to fundamental data.

At this juncture, it is stressed that rolling mill plant, and in particular modern, fully-automated, high-performance mills, can account for up to 40% of the total investment costs of the entire works. Consequently, the greatest care must be taken not only in drawing up the production programme, but also in the selection of the individual mills.

ee) Blooming Mills. Even in very big rolling mills, the blooming mill nearly always represents a potential bottleneck in the chain of equipment, as in most cases no reserve mill is available. It must have a generous capacity, of course, the best basis being the capacity of the steelworks plus 25%. Ilgner sets are still the most popular drive units. Grid-controlled reversing drives are available today with ratings of up to 350 mt. However, savings in space and low investment costs are accompanied by rather high current surges in the works system, and possibly in the public grid. Roller tables, particularly those in heavy mills, nowadays all feature individual motor rollers.

These mills can be of two-high, reversing design (Ilgner, rectifier, or twin drive), or three-high design; roll diameters from 650 mm for small ingots of up to about 1,000 kg up to 1,300 mm for ingots weighing up to 8,000 kg and more. Capacity: above 700,000 t/year if the blooms are not too small (not less than 150×150 mm), and above 2,000,000 t/year when designed as a blooming-slabbing mill. This type of mill is normally of reversing design.

ff) Heavy Section Mills. These mills are used for rolling normal heavy sections above 100 mm in diameter or 100 mm square, and NP 16 and above. Heavy rail mills are usually similar in design to normal medium section mills, and feature heavier rolls, the diameters ranging between 650 to 1,000 mm and more (2—4 stands).

As the material weight per metre is high, these mills have capacities of several hundred thousand tons per year. They can be of two-high design with reversing drives, which saves the rather expensive tilting tables. However, production can be greatly affected if shorter stock (first passes) has to wait until finished material runs out. Even when powerful motors are used for accelerating the rolling speed (with appropriate braking facilities), the production rate still suffers. It is, therefore, better to use either 2 drives for 3 or 4 stands (particularly in mills rolling high-grade and alloy steels), or a three-high mill, which permits rolling independent of the other stands (direction of rotation is always the same).

gg) Medium Section Mills. Mills of this type normally have the following rolling programme: from about 30 mm diameter or square to 100 mm diameter or square, NP 4 to NP 10/14, NP 6 to NP 18 (depending on roll sizes). Roughing stand with rolls of about 500—650 mm in diameter, and 3—4 finishing stands with common drive. Tilting tables. Roll diameter approx. 450—600 mm. Capacity about 50,000—80,000 t/year in two-shift operation with average programme and adequate orders for the various sections.

The capacity can be increased to about 130,000 t/year by deleting the lower dimension range from the programme and providing two drives so that the 4 stands can work simultaneously without interfering with one another.

Fully-continuous medium section mills can be designed for capacities of over 500,000 t/year. However, this calls for very large orders for the various sizes, as programme changes normally mean that all rolls have to be changed. This work can be accelerated by using rapid-change stands; these are made ready for rolling the new sections while the mill is still working with the first set of stands, and replace the latter on the programme being changed. The first set can then be made ready for any further change. This procedure radically cuts change-over times.

hh) Light Section Mills. Rolling programme: from about 5 mm diameter or square to 35 mm diameter or square; flats of up to 300×2.5 mm (skelp) if the mill is designed accordingly, and angles, tees, I-beams and channels up to about NP 6.

Simple mill, non-mechanised, usually comprising a roughing stand with rolls measuring about 475 mm in diameter, and 4—7 two-high interchangeable stands, depending on the lower diameter range.

This type of mill requires a very large operating crew. The absolute investment costs are fairly low, the processing costs high. The mill is easy to operate. Its capacity in two-shift operation is about 10,000 to 30,000 t/year. Outputs of up to 100,000 t/year can be achieved if continuous roughing stands and cross-country stands are used and the finishing stands split into two groups, each with its own drive. The last group operates at a higher speed than the first group.

Multi-stand continuous roughing mills (more than 20 stands, followed by cross-country stands, rod finishing blocks or strip finishing train) with the possibility of rolling in two strands down to about 8 mm for the cooling bed, and with coiling facilities, can reach outputs of up to 300,000 t/year. Skew roller tables should be used to accelerate the rolling of larger sections.

Single-purpose wire-rod mills of fully-continuous design with up to 4 strands and exit speeds of 50 m/s have capacities of up to 700,000 t/year.

ii) Blooming-Slabbing Mills. The planning of combined blooming and slabbing mills demands special care and attention. Mills designed with 2 or 4 vertical rolls are naturally expensive and complicated, and are built only when certain conditions obtain. They are used in cases where a billet mill and a plate or sheet mill have to be supplied with initial material and the number of ingots and slabs involved do not justify an ingot mill and a slabbing mill, or in instances where market fluctuations cause frequent changes in the tonnage ratio between ingots and slabs, and consequently in the utilisation of the billet and plate/sheet mills.

jj) Plate, Sheet, and Strip Mills. Plate mills are characterised by the barrel length of the rolls, which vary between 1,800 and 5,200 mm; the maximum plate width is some 200—300 mm narrower than the barrel length.

Modern heavy and medium plate mills feature a very wide range of equipment, and only a few typical examples can be described below. In earlier times, heavy and medium plate was produced in two-high reversing mills or three-high Lauth mills with one or two stands; either slab ingots or rolled slabs were used as starting material.

Normally, vertical edging stands were used to get rid of the taper on slab ingots and to control the plate width.

Plate Mills. *The most simple plant for rolling heavy and medium plate* consists, for example, of a three-high Lauth stand (top and bottom rolls about 850 mm, middle roll 650 mm in dia., barrel length 2,000 mm). The initial material can take the form of slab ingots or rolled slabs.

Capacity: Up to 60,000 t/y heavy plate in 3-shift operation.

Production programme: Plate gauges of 5—25 mm at widths of 550—1,800 mm.

In order to improve the surface finish, produce closer tolerances, and increase output, modern four-high stands are used to finish plate rolled in three-high Lauth or two-high reversing stands.

Modern plate mills of the simplest design consist of a four-high reversing stand (work roll diameter 900—1,000 mm, back-up roll diameters 1,450—1,600 mm, barrel length up to about 3,400 mm), preceded by a vertical edging stand (roll diameter of about 900—1,000 mm). Starting material: slab ingots 700—1,500 mm in width, 120—250 mm thick, up to 3,200 mm long, and weighing approx. 1.8 to 8 tons.

Rolling programme: Plate 700—3,100 mm in width, 4—40 mm thick, strand lengths of up to about 30 m. Capacity on 3-shift operation: 240,000—360,000 t/y, depending on programme mix.

A higher-capacity mill comprising a four-high reversing roughing stand (barrel length of 3,400 mm, for example), a vertical edging stand, and a four-high reversing finishing stand can produce up to 500,000 t/y heavy plate using slabs weighing up to about 15 tons in three-shift operation. Rolling programme: Heavy plate 500—3,100 mm in width, 3—40 mm thick, strand length of up to 25 m.

Large quantities of thinner material, about 5—12mm thick, can be rolled in semi or fully-continuous wide strip mills with a 4/6-stand four-high finishing train. The coils produced in this type of mill are cut to produce sheets up to 2,200 mm in width.

A new development, *the combination slab-plate mill*, is used for the production of blooms and slabs from ingots and slab ingots; in later shifts, the slabs are rolled down to heavy plate.

The mill comprises soaking pits, a pusher-type furnace, a vertical edging stand, and the rolling mill proper, which can be used as a two-high reversing stand (roll dia 1,068 mm) or as a four-high reversing stand for rolling heavy plate. (Work roll dia 965, back-up roll dia 1,346, barrel length 2,895 mm).

In 24 h, ingots or slab ingots are rolled for 2×4 h, and heavy plate for 2×8 (alternately). 60—80 minutes are required each day for the 4 roll changing operations. Plate programme: Above 6.35 mm to about 40 mm thick, up to 2,600 mm in width, strand length up to about 24 m. This multi-purpose mill opens up the possibility of rolling heavy plate from slab ingots in two heats with a controlled slab surface quality without having to invest in a pure heavy plate mill.

Sheet Mill (Example). Older mills for the production of sheet in the range of about 0.18 to 2.75 mm thick in widths of about 530—1,250mm and lengths of up to 2,500 mm feature 1—4 two-high stands with roll diameters of 700—900 mm. The barrel length governs the sheet width. A 2-stand mill of this type (with a barrel length of 1,200 mm) turns out some 10,000—15,000 t/y in three-shift operation.

Starting material: Sheet bar 180—400 mm in width, 750—1,300 long.

The width of the sheet bar governs the sheet width. The sheet bar thickness (8—45 mm) at a given width is determined by the required sheet weigth or thickness.

Output was increased by adding a three-high (high-speed) Lauth stand (in the same rolling axis) with various facilities as the roughing mill to supply the following (usually 2) two-high stands with initial material. Capacity: up to 40,000 t/y.

Sheet mills rolling *soft commercial steels* have completely lost their significance since the introduction of wide strip mills (followed by cold rolling mills), which turn out an incomparably better product (quality), have a much greater capacity, and a high yield. Mills with a three-high mechanised Lauth stand and 1—2 two-high stands are nowadays used almost exclusively for rolling special materials, e.g. transformer, stainless, heat-resistant, and tool steels down to sheet. These mills remain competitive, having a high degree of flexibility even when orders are small.

Strip Mills

Narrow Strip Mills

Rolling programme: strip 50—300 mm wide, 1.2—5 mm thick.

Starting material: Billets 80 and 100 mm square, and slabs 12—300 mm in width, 80—100 mm thick. Specific coil weights up to about 3 kg/mm strip width.

Semi-continuous mills feature a three-high or two-high stand, rolls 550—650 mm in diameter with 2 vertical edging rolls, arranged behind a continuous train of 4—6 four-high or two-high stands and 2—4 edging stands (four-high stands guarantee excellent gauge tolerances).

Capacity: Approx. 110,000 t/y coils, working in three shifts.

Fully-continuous mills comprise 7 two-high stands (roughing train), four edging stands, and four 4-high stands (finishing train).

Capacity: Up to approx. 200,000 t/y coils.

Medium Strip Mills

Rolling programme: Coils up to about 300—700 mm in width, approx. strip thickness 1.25—8 mm (normally fully-continuous).

Starting material: Slabs 300—700 mm in width, 75—130 mm thick. Specific coil weights up to about 5 kg/mm strip width.

Fully-continuous mills feature 7 two-high stands (roughing train), 5 four-high stands, and upt to 6 edging stands. Capacity: Up to 450,000 t/y working in three shifts.

Wide Strip Mills

These mills can be of *semi or fully-continuous design.*

Rolling programme: Coils 600—2,200 mm in width, strip thickness 1.5—10 mm.

Starting material: Slabs, rolled or continuously cast, 600—2,200 mm in width, 150—300 mm thick. Specific coil weights: 5 to 18 kg/mm strip width.

The strip width, specific coil weight, and strip speed govern the capacity and design of the mill.

Semi-continuous Mills.—Steckel Mill. Semi-continuous wide strip mills comprising an edging stand, a four-high reversing stand, a four-high Steckel mill, and two furnace coilers have the lowest possible annual capacity. (Three to five passes are taken in the Steckel stand). Rolling capacity: Up to about 300,000 t/y wide strip measuring 700—1,200 mm in width and 2—5 mm in thickness, working 3 shifts per day. Specific coil weight: Up to 5 kg/mm strip width.

Some mills feature two/three 4-high finishing stands behind the Steckel stand; this permits an increase in capacity, as only *three* passes are taken in the Steckel stand.

Capacity: Up to 1×10^6 t/y working three shifts.

Higher-capacity Wide Strip Mills. These mills can consist of an edging stand, a four-high reversing stand, and a six-stand four-high finishing train with 2 or 3 coilers.

Capacity: Approx. 1.3×10^6 t/y coils in three-shift operation.

Rolling programme: Strip 700—1,250 mm in width, 2—5 mm thick, specific coil weight 8 kg/mm strip width.

Fully-continuous Mills. These mills usually comprise a roughing and intermediate train of up to 5 stands and a finishing train with 6/7 four-high stands.

Capacity: Approx. 2×10^6 t/y coils rolling strip 600—1,500 mm wide with a specific weight of 10 kg/mm strip width at a maximum exit speed of up to 12 m/sec.

During the past few years, wide strip mills with parallel four-high heavy plate stands for plate widths of up to 3,000 mm have been commissioned; depending on market demands, these mills are used for rolling *either coils or heavy plate.*

Combined Wide Strip and Heavy Plate Mills. The roughing stands of wide strip mills are often used for the production of heavy and medium plate in order to save investment costs; this practice is frequently adopted during expansion stages, for example. Instead of the rolled material being passed on to the actual wide strip mill (Steckel or continuous mills), it is pushed out laterally for finishing.

The diagrams shown below depict a Steckel mill in a combined plant (wide strip—heavy/medium plate), and the separate use of a pure single-purpose wide strip mill and a pure heavy/medium plate

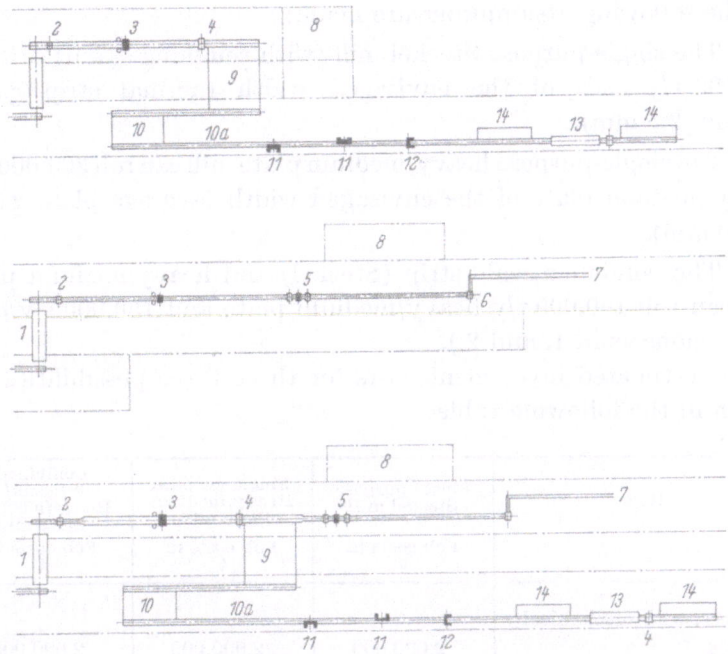

Fig. 25. Medium/heavy plate mill (example).

1 Pusher-type furnace; *2* scale washer; *3* reversing stand with edger 900/1500 dia × 2900 mm; *4* leveller; *9* cooling bank; *10a* marking section; *11* trimming shear; *12* dividing shear; *13* annealing section; *14* piler.

Steckel mill (example).

1 Pusher-type furnace; *2* scale washer; *3* reversing stand with edger 700/1250 dia × 1400 mm; *5* Steckel stand with coiler furnaces 650/1250 dia × 1400 mm; *6* coiler; *7* coil conveyer.

Combination Steckel — heavy/medium plate mill (example).

1 Pusher-type furnace; *2* scale washer; *3* reversing stand with edger 900/1500 dia × 2900 mm; *4 + 4a* leveller, *4a* mobile leveller; *5* Steckel stand with coiler furnaces 650/1250 dia × 1400 mm; *6* coiler; *7* coil conveyer; *8* roll turning shop; *9* cooling bank; *10* inspection and plate turnover gear; *10a* marking section; *11* trimming shear; *12* dividing shear; *13* annealing section; *14* piler.

mill. In this example, it is assumed that the initial material is slabs measuring 1,250 mm in width (either rolled slab ingots or continuously-cast slabs, heated in pusher-type furnaces).

As the sketches show, the roughing stand of the combined mill is heavier than the roughing stand of a single-purpose Steckel mill, as it is used for rolling heavy and medium plate, too.

In order to demonstrate the problems attached to both possibilities, the production rates and investment costs are compared below.

The following assumptions are made:

1. The single-purpose Steckel mill (with roughing stand) can roll 300,000 t/y coils of the envisaged width (normal strip gauge approx. 2.5 mm).

2. The single-purpose heavy/medium plate mill can roll 300,000 t/y heavy/medium plate of the envisaged width (average plate gauge 8—10 mm).

3. The combined wide strip (Steckel) and heavy/medium plate mill can roll 150,000 t/y heavy/medium plate and 150,000 t/y coils (dimensions as in 1. and 2.).

The estimated investment costs for these three possibilities are shown in the following table:

Item	Single-purpose Steckel mill Fob costs in $	Single-purpose Heavy/medium Plate mille Fob costs in $	Combined Steckel/ Heavy/medium Plate mill Fob costs in $
Furnaces, with mechanical facilities	2,000,000	2,000,000	2,000,000
Heavy/medium Plate mill		4,000,000	
Roughing mill	2,000,000		4,000,000
Steckel mill	2,500,000		
Finishing equipment		5,000,000	5,000,000
Electricals	4,000,000	5,000,000	7,500,000
Bays, less stores	4,000,000	5,000,000	6,500,000
Cranes	1,500,000	1,500,000	2,000,000
Total equipment, fob	16,000,000	22,500,000	27,000,000

Even though these fob investment costs are only roughly estimated, they do reveal the following tendency:

1. Single-purpose Steckel Mill: 16,000,000 $/300,000 t/y = approx. 53 $/t coils p. a.

2. Single-purpose Heavy/medium Plate mill: 22,500,000 $/ 300,000 t/y = approx. 75 $/t plate p. a.

3. Combined Strip—Heavy/medium Plate mill: 27,000,000/ 300,000 t/y = approx. 90 $/t coils—medium/heavy plate p. a.

These figures speak for themselves and need not be commented upon. Mention must, however, be made of the fact that where a shortage of funds or the programme mix makes the erection of a combined strip-plate mill necessary, this plant should be so laid out that the two mills can later be operated separately after suitable furnaces and a new stand (as roughing stand for the Steckel mill or heavy/medium plate mill) have been added.

Planetary Mills. Slabs measuring some 60 mm in thickness can be rolled down to approximately 2.0—1.5 mm in hot Sendzimir mills, Platzer mills, and similar facilities, which are used for lower tonnages of strip up to 1,000 mm in width. These mills feature very slender working rolls (approx. 50 mm dia.), which are backed by thicker rolls. However, this type of mill is very complicated and requires expert servicing and maintenance.

kk) Sheet and Strip Finishing Processes. All facilities for the *continuous* processing of coiled hot-rolled strip—e.g. continuous pickling, annealing, tinning, galvanising—are known as *strip processing lines*. The ends of the coils are welded together to produce an endless strip of steel; between the processing lines, the strip is divided to produce coils of various weights for rolling, cleaning, etc.

The further processing of strip and sheet is a very complicated field, and no effort is made here to treat the subject in full; instead, the *principal* features of strip processing lines and discontinuous hot-dip plant for sheet are described.

The time utilisation factor of strip processing lines can be taken at 80—90% of the operating time.

As cold-rolled sheet and strip have superior surface finishes, the treatment of the surface prior to actual processing is of prime importance.

Pickling Plant. Scale resulting from the hot rolling process must be removed from the strip surface in a pickling plant prior to the cold rolling process. For this purpose, the strip is led through tanks, usually ceramic-lined, which can contain sulphuric or hydrochloric acid. The most widespread practice today is to lead the strip *horizontally* through tanks containing *sulphuric* acid. Consumption of the H_2SO_4 is approx. 10—12 kg/t strip (the iron sulphate is regenerated by vacuum crystallisation; this produces iron sulphate-heptahydrate).

Capacity: Up to 1×10^6 t/y strip (reference strip strip of 1,100 × 2.2 mm).

Pickling speed: Approximately 100—180 m/min (larger facilities).

Now that difficulties experienced in the regeneration of the spent solution have been overcome, the tower pickling process (strip is led vertically) using hydrochloric acid is becoming increasingly popular.

Consumption of HCL is approx. 5 kg/t strip (the solution is regenerated by hydrolysis; this produces pure iron oxide).

Either of these two pickling processes can be used, the choice being governed by local conditions, availability and price of the acids, etc. The pickling effect is the same in both cases.

Cold Rolling Mill. After pickling, the coils are put through a cold rolling mill (the heaviest possible coils should be used—up to 100 t).

Here the strip is heavily reduced (up to 92% reduction) in 4-high reversing stands or 3/6-stand tandem mills (soft steels are not subjected to intermediate annealing). Rolling speeds: Up to 1,800 m/min. 1 and 2-stand skin-pass mills are used for the second (minor) reduction (after intermediate annealing).

Selection of cold rolling mills for these lines is governed by the strip quality, the programme mix, the strip gauge and width tolerances, the tonnage to be handled per annum, and various other factors such as the utilisation factor of cold rolling mills (approx. 73%—some 27% is accounted for in unavoidable down-times). For example, a 5-stand tandem mill rolls stock at full operating speed for approx. 48—50% of the actual shift.

Mill design is based on the rolling programme, e.g.:

Strip gauge	Mill design
1.5 mm	Reversing mill
0.7 —2.0 mm	3-stand tandem mill
0.4 —1.25 mm	4-stand tandem mill
0.25—0.50 mm	5-stand tandem mill
0.18—0.40 mm	5-stand tandem mill of heavy design
below 0.18 mm	6-stand tandem mill of heavy design

This rough division does not cover every possibility, of course. For example, it is possible to roll strip with a minimum gauge of 0.18 mm in reversing 4-high mills, taking five passes.

Cold rolling mills for alloyed steels are of special design. Cold Sendzimir stands (20 rolls) and MKW stands (with small work rolls) are particularly suitable for this application. Cold Sendzimir stands feature rolls measuring some 50 mm in diameter, and roll at speeds of up to 300 m/min.

These mills permit the rolling of alloyed steels (e.g. stainless steels) without intermediate annealing in spite of the high degree of cold reduction; the rolled material has an exceptionally good surface quality and extremely close tolerances.

Hood Annealing Furnaces. The cold-rolled coils are stacked on flat bases (stack heights: up to 5 m) and enclosed in a steel cover for annealing at temperatures of about 700—800 °C (inert gas atmosphere). The oil or gas-fired heating hood is then placed over the steel cover (temperature 1,000 °C). Each heating cover hood for some 3—4 bases; i.e. while one base is being used for annealing, a second is being prepared, and either one or two accommodate coils that have been annealed and are cooling off.

Capacity per heating cover: Up to 2.5 t/h. Heat consumption: Up to approx. 200,000 kcal/t. Propane can be used for the production of inert gas.

Continuous Strip Annealing Furnaces. These installations are used for tinplate, which is guided either horizontally or vertically, depending on the design of the plant. Considerable economic advantages are offered, as the output rates are high. The strip has a finer grain structure than strip annealed in cover furnaces; consequently, it has a higher yield point and a higher tensile strength; the deep drawing properties of both materials are virtually the same. The strip is annealed in an inert gas atmosphere.

	Vertical plant	Horizontal plant
Strip speeds	Up to 600 m/min	180 m/min
Capacity	Up to 500,000 t/y	100,000—120,000 t/y
	with a reference strip of 760 × 0.24 mm	
Heat consumption	Up to 130,000 kcal/t (indirect gas firing)	

Skin-Pass Mills. Annealed strip is always put through a skin-pass mill to prevent the formation of stretcher strain marks during subsequent working (deep drawing, etc.). Degree of reduction: about 1—3%.

The mill usually comprises one four-high or two-high stand; tin-plate is normally skin passed in two-stand tandem mills (non-reversing). Present and future demands must be allowed for when planning these mills. The maximum rolling speed is about 1,500 m/min.

For low production rates, reversing four-high stands can be used for cold rolling (with emulsion) and for skin-pass rolling (after removal of all oily residues).

Capacity planning *must* be based on a reference strip.

In recent times, two-high skin-pass stands have been installed behind hot wide strip mills to improve the surface of the strip (descaling—equalising).

Shearing and Slitting Lines for Hot and Cold Strip. These installations operate continuously; the coil ends are not welded together, but are threaded separately into the line.

Shearing Lines. The coils are paid off, the edges trimmed, and the strip divided into sheets; the sheets are levelled, oiled (if required), piled, and made ready for dispatch.

	Speed	Strip gauge
Hot strip	Up to 100 m/min	3—5 mm
	Up to 45 m/min	5—10 mm
Cold strip	Up to 300 m/min	0.3—1.2 mm
Reference width:	1,100 mm	

Utilisation factor: 30—60% of operating time (coil weight and downtimes make themselves felt here).

Slitting Lines. These lines can be operated by themselves or combined with shearing (dividing) lines.

In these lines, wide strip is trimmed, and slit to produce narrow strip; the strip is then coiled. This material can be used for the manufacture of tubes, for example.

Slitting speed: Up to 240 m/min.
Shearing speed: Up to 110 m/min.

Electrolytic Cleaning Lines. Strip *must* be cleaned if it is to receive a metallic coating; cleaning is effected between the first cold rolling operation and the annealing operation, mainly to remove the emulsion, which cannot be completely burned off during the annealing process.

Cleaning is effected in three stages, viz.:

a) Chemical cleaning in an alkaline bath (2—3% solution of Na_4SiO_4); following this, the strip is vigorously brushed.

b) Electrolytic cleaning (the strip is immersed and drawn through 4 sets of electrodes; different sections of the strip act alternately as anodes and cathodes.

c) Spraying, brushing, drying with hot air

Strip speed: 200—600 m/min.

Capacity: About 120,000 t/y.

Reference strip: 760 × 0.24 mm.

Electrolytic cleaning lines can be erected as separate facilities, or combined with continuous annealing facilities.

Separate cleaning plants have simple coil unwinding and recoiling devices. The plant can be stopped for coil changing and for joining coils.

The choice between these two possibilities is governed by the required production rate and the specifications on material structure, etc.

Electrolytic Tinning Lines. There are three basic types of electrolytic tinning lines in use; they are distinguished mainly by the type of electrolyte used:

a) Sulphuric acid plants (e.g. the "Ferrostan" process),

b) Halogen plants,

c) Alkali plants.

Some 68% of the world production of tinned strip is treated in sulphuric acid lines, about 25% in halogen lines, and the remainder in alkali lines.

Process sequence. Cleaning—pickling—electrolytic plating—inductive melting of the tin layer—chemical after-treatment—oiling. All three processes share this sequence.

The "Ferrostan" and alkali processes feature vertical tanks; halogen lines have horizontal tanks.

Only very slight differences are to be noted among the three processes as regards the quality and physical properties of the finished material.

7*

Extensive market research must precede the planning of these facilities (Tinning of coils *or* sheets; differential coating; applications, etc.).

Plant Data (all approximate values)

	Strip speed	Tin thickness in lbs/ basis box	Capacity
Sulphuric acid lines	Up to 500 m/min	0.1 to 1	40,000—180,000 t/y
Halogen lines	Up to 650 m/min	0.1 to 1	200,000—350,000 t/y
Alkali lines	Up to 350 m/min	0.1 to 1	60,000—160,000 t/y

All with a reference strip of: 760×0.24 mm.

All electrolytic tinning lines that feature shears are limited to strip speeds of about 250—350 m/min.

Hot Dip Tinning of Sheet. If tin coatings thicker than 1.25 lbs per basis box are needed, the strip is trimmed and divided in a separate shearing line and the sheets are put through a hot-dip tinning line. *Hot-dip tinning* lines feature an electrolytic pickling tank, the tinning machine (with tin-pot—covered with palm oil), the spraying facility for removing the oil, and dry buffing appliances.

Trimming speed: Approx. 10 m/min

Tin Thickness: Approx. 1—3 lbs/basis box

Capacity: Approx. 8,000—9,000 t/y

Reference strip: 860×0.3 mm.

Continuous Hot-Dip and Electrolytic Galvanising Lines. These galvanising lines are designed for the continuous processing of *coiled strip*.

Hot-Dip Galvanising. In the Armco-Sendzimir process, for example, cold-rolled strip (no intermediate annealing) is cleaned in an alkaline bath, continuously annealed (inert gas atmosphere—recrystallisation) and put through a galvanising pot.

Coilers or shears and pilers are arranged at the exit end of the line.

Strip speed: Up to 100 m/min

Coating: Up to 600 g/m² (both sides)

Capacity: A modern line turns out some 120,000 t/y. Reference strip: $1,100 \times 0.8$ mm.

Electrolytic Galvanising. In this process, coiled *annealed* strip (hood annealing furnaces) is processed in the sequence described below.

Cleaning in an alkaline solution—pickling in sulphuric acid—electrolytic galvanising (zinc anode suspended in solution of H_2SO_4), phosphatising and chromatising (to provide rust resistance and a good base for paints, etc.).

Strip galvanised in Armco-Sendzimir lines has a spangled surface, whereas electrolytically galvanised material has a smooth, matte surface. The latter material lends itself well to shaping and welding, is highly corrosion-resistant, and offers a very good surface for painting, etc.

Sales of electrolytically galvanised strip are now increasing.
Strip speed: Up to 100 m/min
Coating: 36 g/m² (both sides)
Capacity: 60,000—120,000 t/y
Hot-Dip Galvanising of Sheet.
Two main processes are described in brief:

a) *Wet Galvanising.* Following the normal pre-treatment as described above, the sheets are put through a bath of zinc (covered by a fluxing medium).
Galvanising speed: About 3—25 m/min
Coating: About 235—840 g/m² (both sides)
Plant capacity:

b) *Dry Galvanising.* The sheet is led through a bath of zinc chloride (bath temperature about 80 °C). After being dried, the material is put through the galvanising pot (alloyed or unalloyed melt).
Galvanising speed: About 5—40 m/min
Coating: About 180—840 g/m² (both sides)
Plant capacity:

The material flow sheet (Fig. 26) conveys an impression of the basic features of strip processing installations.

ll) Tube Manufacturing Processes (see folding chart). There are two main types of steel tubes:

Seamless tubes (made from ingots, billets and similar materials by piercing and elongating etc.).

Welded tubes (made from strip or skelps; either gas or electric welded—butt or overlapped).

When planning tube manufacturing facilities the following points must be clarified in advance:
Purpose to which the tubes are to be put.

Raw material sources, market conditions, and prices (billets, rolled rounds, continuously-cast billets or rounds, skelp, strip, etc.).

Seamless Tubes. These tubes are made from solid ingots or billets which are pierced to form thick-walled "bottles" or "blooms" in one of the following processes, or from (hollow) cast material:

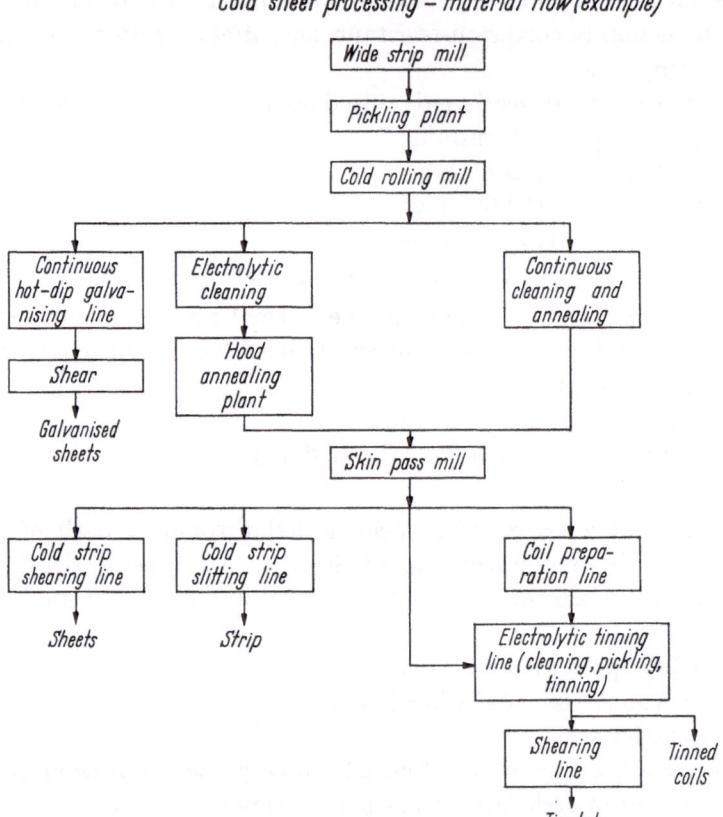

Fig. 26. Cold sheet processing-material flow (example).

Rotary Piercing Process. The solid block is rolled over a mandrel by two barrel-shaped rolls arranged at an angle to the mandrel axis (Mannesmann) or by two mushroom-shaped discs rotating at right angles to the billet axis (Stiefel).

Press Piercing (e.g. Erhardt process). The solid block is placed in a mould and pierced from above by a mandrel. The bottle has a thick bottom.

Cast Material. Cast blocks can be rough-turned and bored, or hollow bodies can be cast and rough-machined both inside and out. Continuously-cast material can also be used.

Elongation Processes. The thick-walled blooms (Mannesmann/ Stiefel) or bottles (hydraulic press—e.g. Erhardt process) can be rolled out to thin-walled tubes in any of the following mills:

Pilger Mills (Rotary Forging Process). The heated bloom is placed on a mandrel and rolled in a series of "jumps" to form a thin-walled tube. The Pilger mill process can be used for the manufacture of tubes with inside diameters of 40 to 550 mm. With normal wall thicknesses (approx. 3—12 mm), tube lengths of up to 30 m can be achieved. Maximum bloom weights: up to 5 tons.

A Pilger mill installation (using billets) for medium tube sizes (60—178 mm) can comprise:
A rotary furnace—cap. 25 t/h
An 800-ton piercing press
A reheating furnace
A cross-rolling mill to elongate the bloom and slightly reduce the walls
Two Pilger stands driven by a common d.c. motor
Two reeling mills (with plugs arranged in the passes) for smoothing the tube surface
One sizing mill for rounding off the tube and producing the required finished diameter.

A stretch reducing mill can be used to manufacture tubes down to 13 mm.

Mill capacity: When manufacturing tubes of 178—13 mm, 70,000 to 100,000 t/y, depending on the programme mix.

This process is eminently suitable for the manufacture of long tubes with wall thicknesses of approx. 3—12 mm. The yield is good. Types of tubes manufactured: API; Special tubes.

Swedish Process (Plug Rolling). The thick-walled blank is rolled down over plugs in 1 or 2 two-high stands with differently sized passes. Several passes are taken to reduce the outer diameter and wall thickness. As a longer mandrel would be liable to buckle, the tube length is restricted to about 6 m. This process has lost its significance; it has a low capacity and a poor yield.

Improved Swedish Process (Plug Rolling). This type of mill is commonly known as the *Automatic Mill,* and is suitable for the production of tubes made of alloy, stainless, and ball-bearing steels in addition to normal grades.

Starting material: Rounds, rough -machined.

This facility can include:

A rotary furnace.

A Stiefel mill for piercing the rounds (with top and bottom guides to prevent undesirable movement in the pass).

The thin-walled bloom is rolled down in a single-stand two-high plug mill; two or three passes are taken, the plugs being replaced by bigger ones at each successive pass to further reduce the wall thickness. The outside diameter remains unaltered. The plug mill is followed by smoothing and sizing stands and the required finishing equipment.

Finished tube lengths of 15 m can be achieved with this process. The Automatic Mill leads the field in the manufacture of API tubes for the oil industry.

Tube (outside) diameters: 60 to 350 mm.

Capacity: Tubes measuring 178—13 mm

approx. 110,000 to 150,000 t/y

356—51 mm

approx. 200,000 to 250,000 t/y

Types of tubes manufactured: Line pipe, API tubes, boiler tubes, special tubes.

Push Bench Processes. Rolled or continuously-cast billets are pierced in a press (ERHARDT) to form thick-walled bottles with a closed end. These bottles are pushed by a mandrel through a series of ring or roller dies (the latter now being the more usual). Each pass is formed by three rollers, offset against one another by 120°. Only one "push" is required to form the tube. Maximum tube lengths: up to 12 m. The wall thickness and O.D. are reduced.

The classical push bench process is attended by difficulties in that uniform wall thicknesses are not easily obtained; this is due to the fact that when the billet is pierced in the hydraulic press a certain degree of eccentricity can occur; this cannot be fully remedied afterwards.

For this reason, modern push bench facilities feature a 3-roll elongating mill (Assel mill) between press and push bench; this permits the use of heavier billets and consequently the manufacture of longer tubes, and removes the eccentricity mentioned above.

Additional information of this book

(The planning of iron and steelworks;978-3-662-28160-4) is provided:

http://Extras.Springer.com

Now that the bottle has fully parallel sides, the walls of the tube are uniform in thickness.

A modern push bench installation comprises:

A rotary furnace
A piercing press
A 3-roll elongating mill (Assel mill)
The push bench
and a reeling mill (mandrel extraction)

A stretch reducing mill can be arranged behind the push bench for the manufacture of small-diameter tubes. Carbon and alloyed steels can also be processed.

Capacity: Tubes measuring 127—10 mm
                        70,000—100,000 t/y
                        178—13 mm
                        100,000 —150,000 t/y

Type of tubes manufactured: Gas pipes, API tubes, special tubes.

Continuous Tube Mills. In this process, rough-turned rounds of approx. 125—175 mm in diameter are used as starting material. The facility comprises: A rotary furnace, two Stiefel mills, and nine two-high stands in which the O.D. and wall thickness of the tube are reduced over a mandrel. Tubes up to about 22 metres in length can be manufactured.

After reheating in a walking-beam furnace, these tubes are rolled down to 13 mm O.D. in a stretch reducing mill with up to 25 stands. Tube lengths: up to 100 m. This is a high-performance mill with a high yield; the tubes are of good quality.

Capacity: Tubes measuring 152—13 mm
                        190,000 to 250,000 t/y.

Types of tubes manufactured: Line pipe, boiler tubes, special tubes.

Elongating Process for Small Tubes Using a Stretch Reducing Mill. The smallest possible mandrel diameter in the Pilgrim, Stiefel, and push bench processes is about 40 mm; consequently, tubes with inside diameters of less than 40 mm cannot be made.

Tubes with inside diameters of less than about 40 mm are rolled in stretch reducing mills with up to 25 stands. Each pass has three rolls, each roll its own drive. Mandrels are *not* used. The tube diameter is reduced, and the wall thickness slightly reduced.

Speed increase from stand to stand is obtained by using gearings or differentials and oil gearings arranged parallel to the mill; this system produces the very high speeds needed in this process.

Stretch reducing mills can be arranged behind Pilger mills, automatic mills, push benches, and tube welding machines.

Depending on the programme, tubes with a standard diameter, for example, 76, 89, 95, or 113.5 mm are put through the stretch reducing mill; by selecting the appropriate stands (which can be cut in and out), a very wide range of finished tubes can be made. Smallest tube diameter: 9.5 mm O.D.

Tube lengths: Up to 200 m

The finished tubes are cut to length by flying saws at an exit speed of 4—6 m/s.

Electric Welded Tubes (Cold Shaped). Continuous narrow strip mills can nowadays produce material with very close tolerances, and great progress has been achieved in the field of electric welding. As a result, the quality of electric welded tubes has been so improved that *seamless and electric welded* tubes are now equally suitable for a wide variety of applications.

Any of the following electric welding processes can be used:
Resistance (butt) welding,
High-frequency resistance welding,
 Induction welding,
Arc welding,
Resistance welding of overlapped edges.

All these processes share two main operations:
The shaping of cold strip or steel plate to form an open tube,
Welding of the seam.

The starting material is prepared by descaling, trimming, scarfing, etc., and the welded tube is finished by deflashing, sizing, straightening, etc.

Not all processes can be covered in detail here, but one is given special mention; this is the Thermatool Process.

The Thermatool Process is a high-frequency (450,000 c/s) welding process which is now being used on an increasing scale for unalloyed steels too; the *skin-effect* (concentration of ac current in the outer layers of solid conductors) restricts the heated zone to a fraction of a millimetre, keeps the weld bead small, and minimises structure alterations. The weld seam is of excellent quality.

A modern welding plant for a wide production programme (tubes measuring 50—350 mm in dia.) comprises:
The strip (or plate) preparation line

The welding plant
A reheating furnace
A stretch reducing mill
Finishing equipment
Possible capacity: 120,000—160,000 t/y

Gas Welded Tubes (Hot Shaped). In the Fretz-Moon process, for example, the starting material is heated for shaping, and butt-welded with gas before being passed into a stretch reducing mill (for diameters below 50 mm) with an output of up to about 35 t/h.

Starting material: Strip, preferably basic Bessemer quality; type of tubes manufactured: gas and water pipes.

Precision Steel Tubes. Precision steel tubes with excellent surface qualities and extremely close tolerances can be made from hot-rolled tubes by cold drawing or cold rolling. These processes are rather involved, and a detailed description of them would exceed the scope of this book.

mm) Rolling Mill Furnaces[1]. In this section, only the principal types of rolling mill furnaces are described. Other (special) furnaces are covered in the section on rolling mills (see page 97).

Soaking Pits. It is obviously a good practice to transfer ingots to the rolling mill furnaces immediately after stripping in order to conserve the heat still contained in them. This applies to heavy ingots in particular; the heat in the still molten interior of a heavy ingot is allowed to penetrate to the outer regions while the ingot stands upright in a soaking pit, which are often arranged ahead of blooming and slabbing mills. In addition, the soaking pit serves as a buffer between the steelworks, which supplies ingots in batches (tap-to-tap time), and the rolling mill, which works with a steady rhythm.

Soaking pits can be designed with separate cells or with one large-capacity pit. If very heavy ingots or slabs have to be soaked, or if special steel grades have to be kept at specific temperatures, new soaking pits will feature separate cells, but otherwise the large-capacity design will be given preference. Generally speaking, the

---

[1] LÜTH, F.: Gesichtspunkte für Bestellung, Abnahme und Beurteilung hüttenmännischer Glüh- und Wärmofen (The Ordering, Inspecting, and Assessing of Annealing and Heating Furnaces). First published internally in 1937 by the Verein Deutscher Eisenhüttenleute. Final version by LÜTH-GUTHMANN published in 1950 by the Verein Deutscher Eisenhüttenleute.

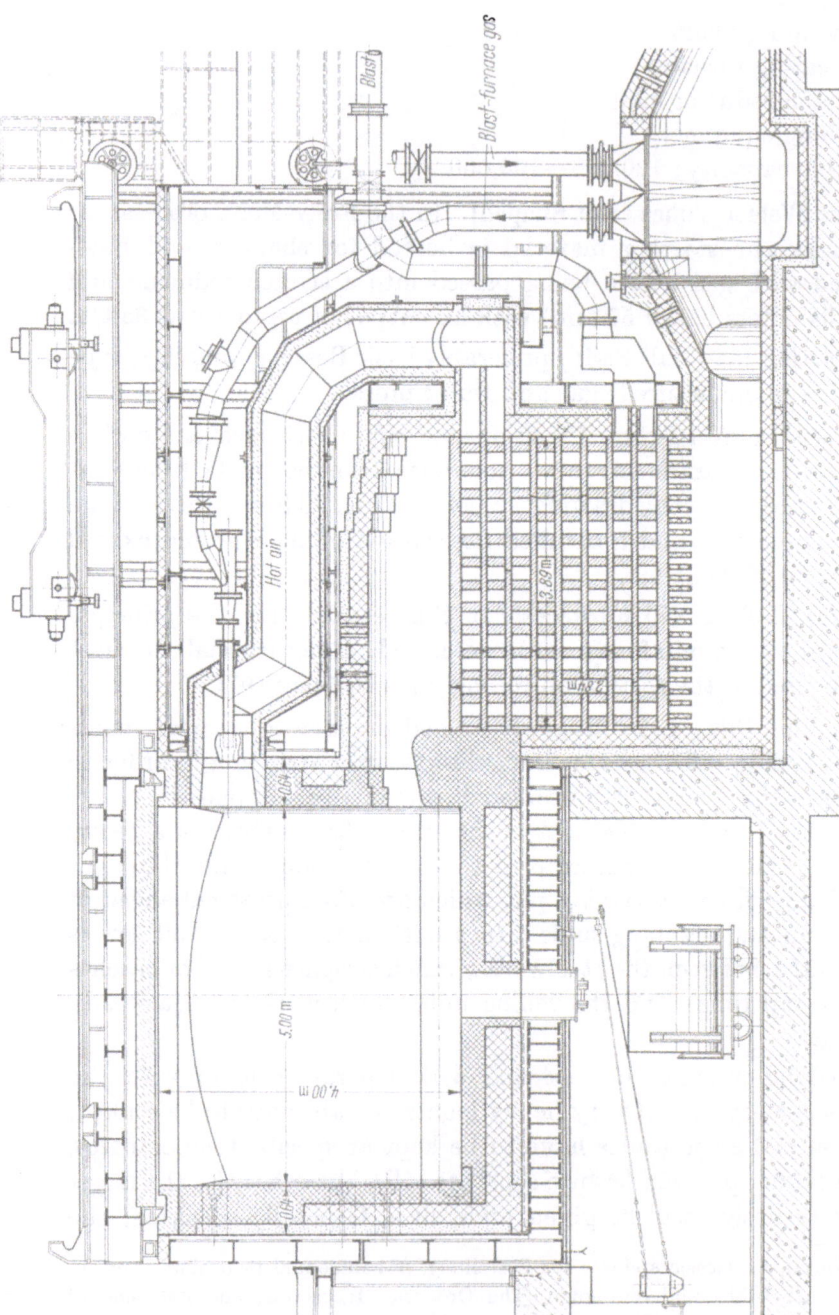

Fig. 27. Large-capacity soaking pit (one-way fired) with a ceramic and a steel recuperator for heating the combustion air and blast-furnace gas.

Hearth area 11.2 m²; air preheated to 1000 °C, gas to 400 °C; heat consumption (cold charge) 380 kcal/kg. Capacity: 5.5 t/h (cold charge), 10—16 t/h (hot charge), soaking times 2 and 4 hours respectively. Holding capacity 10 ingots at 5 t each = 50 t.

orthodox recangular pit is still the most popular; the circular
soaking pit introduced into Europe by the Salem Engineering
Company in 1939 is an exception. This pit features tangentially-
arranged burners, a central gas offtake, and a central drain for
liquid slag. From the heating point of view, this system has several
definite advantages to offer. However, the necessity of having to
repair the lining at the ingot head level at very frequent intervals
is a disadvantage that has braked the further spread of this type

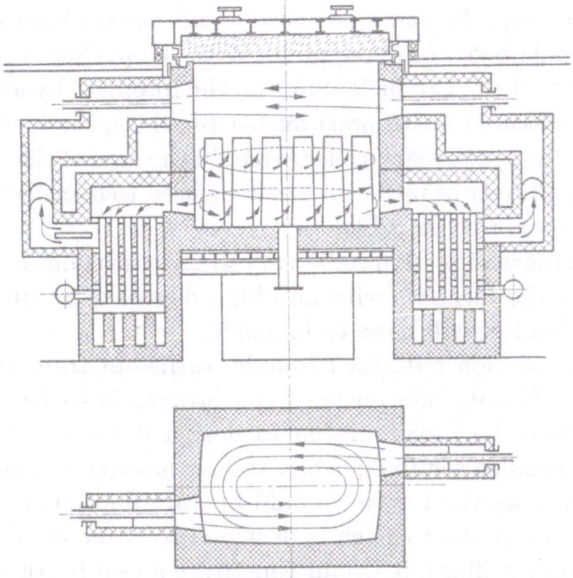

Fig. 28. Section through a two-way fired soaking (rich gas) with a ceramic recuperator for preheating the air.

of soaking pit, at least in Europe. The new types of soaking pits
developed in the USA by AMCO and LOFTUS have achieved great
popularity in Europe during the past few years. The successful
reintroduction of the ceramic recuperator after many years of disuse
is most remarkable; this type of recuperator permits the high-
temperature waste gases of the soaking pit to be put to the fullest
use. The large-capacity soaking pits shown in Figs. 27 and 28 both
feature ceramic recuperators. The pit shown in Fig. 27 is one-way
fired; in addition to the ceramic recuperator for preheating air, it
features a steel recuperator for preheating the blast-furnace gas.
The pit shown in Fig. 28 is two-way fired.

Electrically heated soaking pits have recently gained in popularity where cheap electric power is available, where gas supplies are
insufficient, or simply because of the excellent degree of control
offered.

As the rolling programmes of blooming and slabbing mills are
subject to little or no fluctuation, correct dimensioning of soaking
pits is not difficult. The basis is the ingot supply cycle and the mill
capacity.

Pusher Type Furnaces. As soaking pits call for expensive crane
plant, pusher type furnaces are used in all cases where ingots are
allowed to cool down after stripping. With the pusher type furnace,
the ingots (or billets) can be laid flat on the foreplate by an ordinary
crane; in the case of continuous pusher furnaces, the ingots simply
fall onto the (mill) approach roller table; in the case of side-discharge
furnaces, a pull-over or push-out device is used to transfer the ingots
to the appropriate roller table.

As the outputs of section mills vary greatly, it is not always easy
to dimension the furnaces with any high degree of accuracy; in all
cases, compromises will have to be made.

A medium section mill, for example, turns out from 10 to more
than 30 t/h. If only one pusher type furnace is to be erected, it
would naturally be unfavourable to design it for the full rolling
capacity of smaller sections; when rolling heavier sections with a
lower hourly rate, the furnace would operate uneconomically. On
the other hand, if the furnace is designed to cater for the heavier
sections (slower rolling), it would constitute a real bottleneck when
rolling smaller sections at a higher hourly rate.

The situation calls for careful thought; in many cases, the best
alternative will be to erect two or more furnaces, possibly with
different capacities, even though this involves extra investment
costs and more operators are required.

Today, pusher type furnaces are available in a very wide range
of capacities. Standard single-zone furnaces have a capacity of
about 3—5 t/h; five-zone high-performance furnaces (for slabs)
handle up to 200 t/h.

The furnace rails are usually water-cooled. This can be dis-
advantageous as far as high-grade and alloy steels are concerned;
for this reason, the rails are often clad with heat-resistant steels
(which in turn demands extreme care when using fuels with sulphur
contents).

Walking beam furnaces are often used for this application. The material is not moved by a pusher, but carried along by the beams, which are mounted on eccentric shafts.

Annealing furnaces are mostly used in the manufacture of high-grade and special steels. Cover type annealing furnaces for sheet are described in the section on cold rolling mills. *Continuous annealing furnaces* for tinplate feature heat-resistant rollers for conveying the material. *Box annealing furnaces* using chips as an anti-oxidant are no longer used on a significant scale.

At one time, half-gas generators (coal basis) were used for heating. Today, oil and gas are the most popular heating mediums. The sulphur content of these fuels must be closely watched, especially where high-grade steels are concerned.

If the heating gas has less than about 1,700 kcal/Nm^3, recuperators for gas and air must be used, or oil or rich gas added (natural gas has approx. 10,000 kcal/Nm^3—blast-furnace gas about 1,000 kcal/Nm^3).

3. Forging Hammers and Presses

Even though the majority of forging hammers are found in the metal working industry and not the iron and steel producing industry, many iron and steelworks, in particular those producing high-grade steels and railway material, have their own forges.

Most presses are nowadays of the hydraulic type and exert a force of several thousand tons. Forging hammers are usually steam-driven; during the past few years, higher steam pressures (15 atmospheres and more) have become quite common, and superheated steam is used to combat condensation. The steam is seldom exhausted these days, but collected for recovery of the condensate. During the past 25 years or so, compressed air has to a large extent replaced steam for driving small and medium capacity hammers; most large hammers are still driven by steam. The use of compressed air obviates steam condensation losses (danger of water hammer); in addition, compressed air is cheaper to produce, the plant requires less space and maintenance, and the investment costs are lower. The practice of using recuperators to preheat the compressed air has proved advantageous. Compressed air is used at about the same pressures as for steam; hammers designed for steam operation can easily be converted for use with compressed air.

Forging furnaces are very important items of equipment and exert a high degree of influence on production costs and the running of the forge.

Furnaces for heating large forgings, and in fact all discontinuous furnaces, should be of special design and feature light firebrick linings or similar refractories; in other words, the furnaces should feature a minimum of heat-storing masses[1].

III. Ancillary and Auxiliary Departments

The ancillary and auxiliary departments of an iron and steel-works account for $\frac{1}{4}$ to $\frac{1}{3}$ of the total investment costs for the entire works. In this section, only the most important of these departments are discussed.

See Section D. I, page 165 for a detailed description; the example given clearly indicates the scope of the various departments.

a) Power and Energy

1. Power Station

Exceptional cases apart, the building of an adequate works-owned power station is always desirable, if only to relieve the public power system; this also applies even when it is not intended that power be generated for the grid too, under normal circumstances. A works-owned power station is *imperative* for all iron and steelworks that have no surplus gas.

This applies to works that have no coking plant and to those with a long list of heavy consumers.

The planning of even the largest works-owned power station must be based on the fact that the power station is always an *auxiliary,* and that the question of profitability (steam and fuel consumption) ranks second to the demand for 100% *operational dependability.* It may be necessary to leave out reheaters and multi-casing turbines, using single-casing turbines without reheaters, even though this may increase steam consumption by 1 or 2%. The same applies to steam pressures. Whereas modern high-capacity power stations use pressures of 120 to above 200 atmospheres, a power station of equal size for an iron and steelworks is best operated at steam pressures of 80 to a maximum of 100 atmospheres. Of

---

[1] See E. SENFTER: Feuerfeste Isolierbausteine als Baustoffe neuzeitlicher Glüh-öfen (Refractory Insulating Bricks for Modern Annealing Furnaces). Arch. Eisen-hüttenw. 8 (1934/35) pp. 473/78 (Mitt. Wärmestelle 215).

course, all boilers, turbines, pumps, etc. must be standardised. The boiler control system must be comprehensive and permit control of the entire power station from *one central point*. This is most important if the demand for immediate response (gas peaks, load peaks) is to be met.

The question as to which *solid fuels* should be used, e.g. lignite, hard coal (as fines, lump coal, coal dust, etc.), is usually answered by availability; the boilers must be designed for the particular type of coal available to the iron and steelworks. Depending on the volume of blast-furnace gas (and gas peaks) to be expected, all or some of the boilers must be designed for firing with blast-furnace or coke-oven gas; they should be designed for gas *and* gas/coal firing. The gas volumes are established during the preliminary planning stage.

If at all possible, *all* boilers should be designed for firing with blast-furnace gas, coke-oven gas, or coal—together and singly— in order that all future demands can be catered for. This will be relatively easy to accomplish where coal dust is fired; where grate-firing is employed, things will be much more difficult. Attention is also drawn to the fact that with blast-furnace gas firing, considerable variations take place in the heat transfer conditions between the combustion chamber and the gas ducts; when switching from coal dust to blast-furnace gas, this greatly affects conditions in the superheaters in particular.

Power stations are seldom planned to cater for the full power requirement of the iron and steelworks; for obvious reasons, the power station should and must be coupled with the public power supply system. For this reason, the works power station is normally planned on the basis of the maximum available volume of surplus blast-furnaces gas. In isolated cases, works-owned power stations are planned as pure emergency facilities. Where a link-up with the public power supply system is not possible—and this will be restricted to a very limited number of cases—the power station must be designed not only to cater for the full power requirement of the iron and steelworks, but also to have a reserve capacity of about 25%.

Assuming that a modern, integrated iron and steelworks consumes 250 to 350 kWh per ton of crude steel, the following values would apply for a steel plant with an annual capacity of one million tons:

8 Lüth/König, Steelworks, 3rd Ed.

Consumption rate kWh/t crude steel 250 to 350, average 300
Average power requirement p.a.   kWh/y   300,000
Average user hours   h/y   5,000
Average power station capacity   MW   60

Power Station Capacity

a) As an emergency station   10 to 15 MW
b) To use up surplus blast-furnace gas   approx. 25 MW
c) To cover the full power requirement   approx. 60 MW
d) To cover the full power requirement, with a reserve
   of 25%   approx. 75 MW

2. Power Distribution

The use of *standardised* voltages is a matter of course nowadays. Voltages of 220 and 380 are the most common on the low-tension side. However, motors rated at 50 or 100 kW and more are designed for 6,000 kV, of course. Distribution to consumer points through 30 kW ring cables is also a common practice these days. Finally, the laying of the numerous high and low-tension cables in suitable ducts, preferably large enough to allow a man to pass through, is no longer a costly and unimportant "extra", but an efficient feature of modern power distribution systems.

Linking with the public supply system obviates the need to provide reserve capacity, which is a costly business, and also provides the public power suppliers with a very welcome reinforcement

Table 6. *Example of the power distribution in an iron and steelworks with a capacity of 3 million tons p.a.*

(660,000 t coke, 2.4 million t pig iron, 3.0 million t crude steel, 2,315 million t finished rolled products[a])

Department	kWh/t[b]	kWh/t Crude steel	10^6 kWh/y	%
Coking plant	15	3	10	1
Blast furnaces and sintering plant	113	90	271	26
Steelworks[c]	50	50	150	15
Rolling mills	113	87	262	26
Power station (own requirement)	—	25	75	7
Ancillary and auxiliary departments	—	84	252	25
Total/Average	—	339	1020	100

[a] including 6% light sections, 31% structurals, 13% heavy plate, 34% hot wide strip, and 16% cold strip.

[b] per ton of coke, pig iron, crude steel etc.

[c] 12% basic Bessemer, 66% oxygen, 20% OH, and 2% electric steel (crude).

in times of need. In the case of medium and high-capacity power stations, (from about 80,000 kW and more), connection should be made to the 220 kW bus bar, or at least to the 110 kV grid.

Table 6 shows the actual power distribution in an integrated iron and steelworks.

3. Fuels

The overall fuel consumption of integrated iron and steelworks —i.e. the total amount of fuel used less the effective heat produced— lies between 5 and 10 million kcal per ton of crude steel. The large bracket is accounted for by the greatly varying structures of the individual works; the main factor is the difference in the tonnage of pig iron produced. For example, some works have an extremely high tonnage, either for sale or to cover a high proportion of basic Bessemer or oxygen steel in the total crude steel tonnage.

Table 7 shows the fuel balance of a high-capacity integrated iron and steelworks, and is subdivided to show the various fuels, the amounts used, and the amounts produced.

The total consumption per ton of crude steel is 8.27 million kcal/t without coke-oven gas (from the coking plant) and about 11 million kcal/t with this gas.

The gas supply system of an iron and steelworks can only be viewed in its entirety; this means that *all* heat consumers in the works must be included, whether they use gaseous, solid, or liquid fuels. Iron and steelworks are normally supplied with fuels in the form of coking coal, boiler coal, long-distance gas (blast-furnace or coke-oven gas), and oil. If no coking plant exists, the place of coking coal is taken by blast-furnace coke and a certain quantity of small coke.

Table 8 lists in detail the fuel requirements of modern iron and steelworks.

Coke-ovens and blast furnaces, and to a much lesser extent gas producers, are the *gas-suppliers* for an iron and steelworks. Depending on the origin and quality of the coking coal, some 300 to 350 Nm^3 of dry gas are yielded per ton of dry coking coal charged. This gas yield is paired with a gas consumption rate of some 500 to 600 kcal/kg of damp coal; in modern batteries with cross-regenerative type coke ovens, this is covered by blast-furnace gas, coke generator gas, or some other lean gas, or by mixed coke-oven and blast-furnace gas, or by rich gas (500 kcal/kg).

8*

Table 7. *Fuel distribution in an integrated iron and steelworks*[a] (All values in 10⁹ kcal/y unless otherwise stated)

Description	Coal	Coke	Fuel oil	Blast Furnace gas	Rich gas (coke-oven and natural gas)			Total Fuels	Total (all fuels) per t steel		
					1	2	3		10³ kcal	%	%
a) Coking plant	—	—	—	624	61	—	61	685	228	5.2	8.7
b) Blast furnaces [b]	—	5,359	—	—	—	—	—	5,359	1,786	40.4	(67.8)
metallurgical coke other consumers	9	782	8	1,538	453	12	465	2,802	934	21.1	35.5
Blast furnaces, total	9	6,141	8	1,538	453	12	465	8,161	2,720	61.5	(103.3)
c) Steelworks, total	1	85	343	64	6	216	222	715	238	5.4	9.0
d) Rolling mills	—	3	—	724	71	766	837	1,564	521	11.8	19.8
e) Power station	—	—	965	1,111	—	—	—	2,076	692	15.7	26.3
f) Other consumers	—	16	—	38	4	—	4	58	19	0.4	0.7
g) Total consumption a—f	10	6,245	1,316	4,099[c]	595	994	1,589	13,159	4,420	100.0	(167.8)
ditto, %	0.1	47.1	9.9	30.9	4.5	7.5	12.0	100			
h) Total *without* metallurgical coke	10	886	1,316	4,099	595	994	1,589	7,900	2,634	59.6	100.0
ditto, %	0.1	11.2	16.7	51.9	7.5	12.6	20.1	100			

[a] *Capacites*: 660,000 t coke, 2.4 million t pig iron, 3.0 million t crude steel, (60% oxygen, 18% basic Bessemer, 20% OH, and 2% electric steel), total rolling mill capacity (including double-rolled products) 3.36 million t. Power station capacity 40 MW.
[b] The blast-furnace gas yield is deducted from the metallurgical coke; in addition to the blast furnace plant, the sintering plant, gas turbo-blowers, etc. are included.   [c] Useful volume of blast-furnace gas, i.e. the gas volume less 6.5% (losses).

[1] Enriching gas.   [2] For direct heating.   [3] Total of 1 and 2.

The blast-furnace gas analysis depends on the condition and origin of the coke, and on the blast-furnace practice in the works in question; the same factors affect the gas yield, which lies between 3,500 and 4,200 Nm³/t blast-furnace coke.

Blast-furnace gas is rarely distributed at the pressure set by the blast furnace; usually, the pressure obtaining behind the gas cleaning plant is used (with a hydrostatic head of 40 to 300 mm). H. A. BRASSERT introduced into the Salzgitter works the American system of running blast furnaces with a pressure of about 1,000 mm water column (hydrostatic head) in the stack and not compressing the gas any further on its leaving the blast furnace; this practice has given good results. The gas distribution system can feature a gasometer, which functions as a pressure equaliser or regulator rather than as a storage unit.

The distribution of coke-oven gas is governed by the way in which the plant itself is operated. If the coking plant has to supply long-distance gas at the now usual pressure of up to 50 atmospheres, the question of using a high-pressure desulphurisation and by-product plant should be given careful consideration where modern installations are concerned. This offers the advantage of greatly improved control of the desulphurising phase, and the apparatus used is smaller, less expensive, and features much less steel than orthodox facilities.

Blast-furnace gas need not be cleaned and desulphurised for use within the iron and steelworks, but the situation can be different with coke-oven gas. If more than 50% of the total volume of coke-oven gas is distributed as cleaned long-distance gas, the *full volume* should be cleaned; if a high-pressure by-product plant is available, it can also be used to compress this long-distance gas. The extra costs involved are compensated for by not having to deal with two gas systems, two gas pressures, and two degrees of purity.

The best practice to adopt in modern iron and steelworks is to sell as much of the high-grade coke-oven gas as can be spared, using blast-furnace gas and other lower-grade fuels for the works as such to the fullest extent possible. For this reason, most heating and annealing furnaces in rolling mills, forges, and so on feature recuperators for preheating air *and* gas. This permits the firing of either mixed gas or blast-furnace gas, as required.

Table 8. *Heat requirement values*

Plant	Item	Fuel	Heat requirement in			Remarks
			kcal/kg	Nm³/t	kg/t	
Coking plant						
Coke-ovens	dry coking coal	blast-furnace gas	600	600	—	Net calorific value = 1,000 kcal/Nm³
	dry coking coal	coke-oven gas	500	125	—	Net calorific value = 4,000 kcal/Nm³
By-product plant						
Ore preparation						
Sintering strand	dry coking coal	steam	—	—	100—150	Net calorific value = 6,000 kcal/kg
	crude ore	coke breeze +	480—600	—	80—100	
		blast furnace gas[a]	25	25	—	[a] ignition gas
Lurgi furnaces	crude ore	blast furnace gas	400	400	—	
Blast furnaces	basic pig iron	blast furnace coke	—	—	600	Swedish ore
	basic pig iron	blast furnace coke	—	—	900	lean ore
	steel pig iron	blast furnace coke	—	—	650	
	foundry pig iron	blast furnace coke	—	—	800	
	haematite pig iron	blast furnace coke	—	—	800	
	spiegel pig iron	blast furnace coke	—	—	1,200	
	Fe-Mn	blast furnace coke	—	—	2,000	
Blast heating	blast furnace coke	blast furnace gas	800	800[b]	—	[b] net calorific value = 1,000 kcal/Nm³
Blast	blast furnace coke	blast furnace gas	400	400[b]	—	

Lime kilns	limestone	coke breeze gas[c]	650	—	100
OH furnaces	sound ingots	fuel oil + coke-oven gas/natural gas	550	550[d]	—
			1,000—1,200	—	—
Rolling mill					
Soaking pits	charge	gas[e]	150—300	5	—
Pusher-type furnaces	charge	gas	350—700	5	—
Heating furnaces (discontinuous)	charge	gas	350—700	5	—
Open-type annealing furnaces	charge	gas	700	5	—
Continuous annealing furnaces	charge	gas	600	5	—
Hood annealing furnaces	charge	gas	1,000	5	—
Pusher-type furnaces	charge	coal[f]	420—700	—	60—100
Heating furnaces (discontinuous)	charge	coal	420—840	—	60—120
Open annealing furnaces	charge	coal	900	—	130
Box annealing furnaces	charge	coal	1,200	—	170

[c] lean gas with up to 1,500 kcal/Nm³

[d] related to a net calorific value = 1,000 kcal/Nm³

[e] mixed gas and rich gas; in the case of pure blast-furnace gas, an increase of 10% can be expected

[f] Hard coal with a net calorific value of 7,000 kcal/kg

The values of the various fields must be carefully compared and evaluated. Not only must the most economical fuel be selected for a particular application, but the available fuels may have to be taken as a basis for the design (or selection) of the furnaces, etc.[1].

Control and management of the gas system of any iron and steel-works must always be seen as forming a part of the *overall heat economy of the entire works*; this is a prerequisite for the smooth coordination of such things as using peak volumes of blast-furnace gas instead of coal (in the power station) and the substituting of blast-furnace gas for long-distance gas when the demand for the latter is at peak.

In big works, this coordination must be placed in the hands of *one* department, i.e. a duly authorised *energy department*. All relevant measuring and control equipment should be centralised to ensure maximum control efficiency.

As mentioned in the section on blast furnaces (B.III a) 3), developments in Western Europe over the past 15 years have led to the smelting of lean indigenous ores such as mined in the Minette and Salzgitter fields being heavily reduced, or stopped altogether, in favour of high-grade imported ores with 60 to 67% Fe. At the same time, physical preparation of burden material has received very close attention. As a result, not only have coke rates been appreciably lowered, but intensive indirect reduction of the ore in the blast furnace has considerably decreased the calorific value of the gas produced in the blast furnace process (see Table 2, page 23). Due to this reduced calorific value, the gas could not be used for a number of applications without being enriched with other fuels such as light fuel oil, natural gas, or coke-oven gas. Table 9 shows a practical calculation for an iron and steelworks in which the sinking calorific value of the blast-furnace gas as expected during the coming few years is balanced by adding a rich gas (in this case natural gas). By this means, the blast-furnace gas is enriched to produce a net calorific value of 1,100 kcal/Nm^3 for the Cowper stoves, and 950 kcal/Nm^3 for the rolling mill furnaces and similar consumers. Surplus blast-furnace gas is to be fired with heavy fuel oil to heat the power station boilers. In the example given below, the change in blast-furnace gas distribution resulting from the

---

[1] LÜTH, F.: Bewertung verschiedener Brennstoffe (The Evaluation of Various Fuels). Stahl u. Eisen 71 (1951) pp. 327/34.

Table 9. *Example of the need to enrich blast-furnace gas at low coke rates accompanied by low (gas) calorific values*

	Unit[a]	1963	1966	1970
Blast-furnace gas calorific value	kcal/Nm³	950	875	825
Required calorific value	kcal/Nm³	950	1,100	1,100
a) *Gas requirement for Cowper stoves + gas blowers*	10⁶ kcal	65,300	136,000	162,000
Covered by				
Blast-furnace gas	%	100	77	72
Natural gas[b]	%	0	23	28
or blast-furnace gas	10⁶ kcal	65,300	104,700	116,600
natural gas	10⁶ kcal	—	31,300	45,400
Total:	10⁶ kcal	65,300	136,000	162,000
b) *Gas for rolling mills etc.*	10⁶ kcal	107,700	137,000	137,000
Actual calorific value	kcal/Nm³	950	875	825
Required calorific value	kcal/Nm³	950	950	950
Covered by				
Blast-furnace gas	%	100	91	85
Natural gas	%	0	9	15
or blast-furnace gas	10⁶ kcal	107,700	124,700	116,500
natural gas	10⁶ kcal	—	12,300	20,500
Total:	10⁶ kcal	107,700	137,000	137,000
c) *Power station* (boiler house) Original volume of blast-furnace gas[c]	10⁶ kcal	61,200	19,100	43,000
Blast-furnace gas available with natural gas from a) and b)	10⁶ kcal	—	43,600	65,900
Available to power station	10⁶ kcal	61,200	62,700	108,900
Total volume of blast-furnace gas	10⁶ kcal	234,200	292,100	342,000
volume of natural gas	10⁶ kcal	—	43,600	65,900
Total:	10⁶ kcal	234,200	335,700	407,900

[a] All volumes relate to a period of 1 month.
[b] Net calorific value = 7,600 kcal/Nm³.
[c] Calculated *without* natural gas enrichment.

addition of natural gas is expressed in percentage values (for 1970):

	Pure blast furnace gas %	Blast furnace gas %	Natural gas %	Total %
Blast heating	47.4	34.1	13.3	47.4
Rolling mill furnaces etc.	40.0	34.0	6.0	40.0
Power station boilers	12.6	12.6	—	12.6
Ditto, released by addition		(31.9)		
of natural gas	—	19.3	—	19.3
Total:	100.0	100.0	19.3	119.3

Enrichment of the gas used for the Cowper stoves and the rolling mill furnaces cuts the volume of blast-furnace gas used for this purpose by $\frac{1}{3}$, and the gas used in the boiler house increases from 12.6 to about 32%, i.e. is more than doubled by the replacement of blast-furnace gas with natural gas.

In many cases it may prove advantageous to enrich the full volume of blast furnace gas with natural gas to produce a uniform and higher calorific value.

b) Water Supplies

Water supplies have gradually decreased in Germany and other European countries over the past years. As a result, all industrial concerns, particularly iron and steelworks, who consume vast quantities of water, are obliged to plan their water supplies very carefully.

Quantities. The importance of water in iron and steelworks is often underestimated. The uninitiated are usually quite unable to imagine just how much water is needed for cooling purposes, for drinking, and for sanitary installations, etc.

The following rough estimates convey some idea of the quantities of water needed in the various departments.

Coking plant. 5 to 20 m³ per ton of dry coal.

Blast furnaces. 25 to 50 m³ per ton of pig iron; in many cases, much more is required.

This volume applies to *circulation* water; if fresh water is used, the temperature is kept lower and the required quantity is appropriately reduced.

Ore prepartion. It is extremely difficult, if not impossible, to quote water requirement figures in this instance, as ore preparation plants differ greatly from case to case. Cooling water is needed for calcining, sintering, and pelletising plant only, and 200 litres per ton of treated ore can be taken as a guide. Wet ore preparation plants need much more water; the Studiengesellschaft für Doggererze wet plant, for example, consumes 3 to 4 m³/t of crude ore. The greater part of this volume—about 80%—is circulated; the remainder is lost with the tailings and must be made up (see also page 28).

Lime kilns. Of modern design incorporate cooling elements; they require about 5 m³/t of raw limestone.

Basic Bessemer and oxygen steelworks. A consumption rate of 1 to 10 Nm³/t of crude steel can be estimated.

OH steelworks. Open-hearth furnaces need between 10 and 20 m³ of water per ton of crude steel for the furnace end, door frame, and similar cooling elements. In many cases, the volume of water required justifies the use of a re-cooling plant; the water can then be circulated.

Electric steelworks. A consumption rate of 10 m³/t of crude steel can be estimated.

Rolling mills. Between 1 and 20 m³/t of rolled material can be estimated for the various mills (blooming mills to light-section mills) for cooling the rolls, for spray water, and for the cooling channels and rails, etc. of the various heating furnaces. Water consumption is influenced to a high degree by the design of the rolling mills; modern plant will in many cases demand substantially less water than indicated above.

Power station and steam boilers. The generation of steam calls for good feed water, which must be carefully conditioned for use in modern steam generation plant. This is particularly true in the case of high-pressure steam plant. Condensate should be reclaimed for further use. This also applies to condensate from piston-type steam engines, even though the de-oiling process is not always easy to carry out.

If practically all condensate is reclaimed, 90 to 95% of the original volume of water is retained; in other words, only 5 to 10% of fresh water need be conditioned and added to the system.

Some 60 kg of circulating cooling water are needed per kg of steam condensed in condensing turbine plant. Table 10 gives the water balance of an iron and steelworks with an annual capacity of 2.4 million tons of crude steel.

Water circulation and cooling. Scarcely any iron and steelworks has unlimited supplies of such good service and cooling water at

Table 10. *Water balance (example) for an integrated iron and steelworks with an annual capacity of 2.4 millions tons*

Department	Production			Total water consumption[b]			Make-up water		
	Item	Quantity t/t crude steel	m³/t	m³/t crude steel	m³/h[c]	%	as % of the total volume of circulation water[c]	in m³/h	as %
Coking plant	coke	0.75	20	15.00	4,100	8.7	20	820	10.9
Sintering plant	sinter	0.975	10	9.75	2,700	5.7	20	540	7.2
Blast furnaces	pig iron	0.875	40	35.00	9,600	20.3	15	1,440	19.2
Oxygen steelworks	crude steel	1.00	10	10.00	2,700	5.7	20	540	7.2
Rolling mills	rolled steel	0.75	1	0.75	200	0.4	50	100	1.3
Power station[e]	steam	1.30	70	91.00	24,700	52.4	10	2,470	32.9
Ancillaries	crude steel	1.00	11.5	11.50	3,200	6.8	50	1,600	21.3
Totals:	—	—	—	173.00	47,200	100.0	15.9	7,510	100.0

a The works has a coking plant designed to supply the full amount of coke required, only one top-blowing oxygen steelworks, and a power station designed to supply the fully amount of power required in the works.

b Total volume of circulating water.

c These water volumes include a generous safety margin; the amount of make-up water needed can be reduced.

d Related to 8,760 h/y.

e With a power consumption of 350 kWh/t crude steel and a steam consumption rate of 3.75 kg/kWh, the steam volume equals 1.3 t/t crude steel.

its disposal that *no* circulation water need be used. Where this may prove possible, however, the water supply system is simplified in many ways. The water intake temperatures in condensers and cooling elements are lower; in turn, the vacuum in condensers is improved and the amount of steam required per kWh reduced; the same cooling effect is obtained with a lesser quantity of water; cooling surfaces and cooling elements can be reduced in size; and many other advantages of this nature can be exploited. However, the quality of the water is of paramount importance; one question that must be examined is whether the costs incurred for the cleaning, softening, and further treatment of the water cancel out the advantages described above.

Circulation water must always be re-cooled; all departments that use substantial quantities of water—the power station, blast-furnace plant, possibly the steelworks — will need water re-cooling facilities; the remaining departments very seldom require such equipment (only when water is in extremely short supply). Without going into the design details of the cooling plant—cooling towers—mention is made of the traditional wooden structures, and the newer concrete towers used in the USA, which have in recent times gained in popularity in Europe. These concrete cooling towers are oval in cross-section and are of venturi design; they have a longer service life, improved "lift" and therefore better cooling characteristics.

Normal losses—leakages, spray water, etc.—amount to about 5—12% of the total volume of water in circulation; this amount must be made up with fresh water.

At best, the temperature of the re-cooled water lies about 10 °C above the ambient air temperature; because of the formation of boiler scale and other incrustations, cooling water is usually not allowed to exceed a temperature of 50 °C in the cooling elements. In special cases, open hearth furnace cooling water, for example, has been allowed to reach a temperature of up to 60 °C and then used in central-heating systems.

In general, however, further increases in the outflow temperature of cooling water are made impossible by the vitally important cooling requirements of blast-furnace and other plant.

For this reason, it is impossible to utilise the heat in the cooling water flowing from a blast furnace, for instance, for heating purposes, etc., a suggestion often put forward by heating engineers who are very capable in their own sphere, but know nothing of

iron and steelworks practice. At the other end of the scale, excessively low cooling water temperature rises—often only 2—5 °C—
ought to be avoided. This practice does *not* improve cooling, but
merely multiplies the water consumption rate. One argument put
forward for this cooling method is that the cooling boxes are wrongly
designed from the hydrodynamic aspect, and that a higher flow
rate is needed to wash sludge deposits out of the dead corners, so
preventing otherwise unavoidable cooling-box breakouts!

Hot cooling systems will achieve great significance as time goes
on. They consist of tubular cooling elements through which water
is pumped at high pressure (La-Mont). Some plants operate with
pressures of 50 atmospheres. Water and steam are separated in a
collector. Hot cooling has proved its efficiency as a door frame and
furnace end cooling system in OH furnaces; working in conjunction
with a waste-heat boiler, the system produces up to 0.7 t steam
per ton of crude steel[1]. With unlined hot-blast cupola furnaces (see
page 67), hot cooling has given excellent results in place of the
orthodox system of spraying the outer shell. In the foreseeable future, the hot cooling system may well be applied to blast furnaces; this
could cut water consumption to a fraction of the present-day level.

Water quality (water treatment and cleaning): The use of cooling
water as the "best refractory material" in iron and steelworks calls
not only for absolutely dependable water supplies, but also for
a degree of water purity that excludes virtually every possibility
of trouble. This includes mechanical purity, i.e. the removal of
foreign bodies by means of strainers and filters. In some areas,
the water is so hard that brief overheating of individual cooling
elements makes the formation of boiler scale at least *possible*,
even though this is not *probable*. Consequently, service water often
has to be softened before being used.

Water cleaning and treatment facilities are standard items of
equipment and need not be described here.

Water sources. The huge quantities of water needed by an iron

---

[1] See A. HARNISCH: Heißkühlung für Siemens-Martin-Öfen (Hot Cooling for
Open Hearth Furnaces). Stahl und Eisen 73 (1953) pp. 1026/28. — LARDY, H.: Was-
sereinsparung an Kühlvorrichtungen für Siemens-Martin-Öfen (Cooling Water Re-
ductions in OH Furnace Cooling Elements). Stahl und Eisen 73 (1953) pp. 1028/30.
— POPPE, K. E.: Hochdruckdampf aus Abgas- und Kühlwasserwärme von Siemens-
Martin-Öfen (Generation of High-pressure Steam Using the Heat of OH Furnace
Waste Gases and Cooling Water). Stahl und Eisen 73 (1953) pp. 1030/35.

and steelworks (see Table 10) often raise serious problems as regards sources of supply.

Only very big rivers can supply the *full quantity* of water required if no *re-cooling* is carried out, and big rivers are not found everywhere.

In all other cases, the possibility of obtaining water from the following sources must be examined:

ground water,   springs,   rivers,
canals,   lakes,   dams.

Usually, *springs* are not very generous sources of water, and several must be used to provide the quantity of water needed *plus* a safety margin.

Ground water supplies are particularly rich where old river courses with deep gravel beds exist, e.g. the original Elbe valley in the northeast of the Harz Mountains, between Oschersleben and Börssum. Experienced geologists should always be called in to check the conditions and give an expert opinion; at all costs, the danger of the water table dropping in the near vicinity and surrounding area must be estimated and given due consideration.

Rivers provide the simplest and most dependable sources of water, provided they carry sufficient water to permit the quantity required by the works to be tapped off. Dam-regulated rivers (e.g. the Ruhr) are the best of all.

In special cases, water may be taken from *canals*. Naturally enough, the water must be returned to the canal again. This does not affect shipping; in fact, canal engineers are in favour of the weak current obtained by this process. However, as fairly large numbers of people live on canals, the water is subject to pollution of a sort that demands special water treatment processes. Nevertheless, under certain circumstances canal water may have to be used, and the more complicated and expensive water treatment costs taken into the bargain.

Natural lakes and dams are, of course, ideal sources of water for industrial plant.

Quite understandably, all pumping stations and the distributing network of pipes etc. must feature adequate safety facilities and reserve capacities; at least *double* piping systems should be used, preferably independent systems. The water sources themselves should also offer an adequate reserve, the minimum being 50%, the optimum 100% of the peak quantity required.

Drinking water supplies must be planned as a *separate* item. It can be estimated that the drinking water requirement in an iron and steelworks equals about 15 to 20% of the *fresh* service water consumption rate. Normally, iron and steelworks' drinking water systems are linked with those of neighbouring housing estates.

Water distribution within the works. This question is equal in importance to the actual obtaining and conveying of water to the works. Ring mains for each major consumer — power station, blast furnace plant, steelworks, rolling mill—are as much a matter of course as ring mains within the individual departments and for the entire iron and steelworks as such. Reserve pumps with diesel engine or other suitable drive units must also be provided to feed water to the power station, blast furnace plant, etc. in times of need. The drinking water supply system to the actual consumer points must also be laid and operated separately.

In modern iron and steelworks, water distribution and consumption is monitored by up-to-date measuring facilities.

Waste water (volume, cleaning, disposal). The question of waste water disposal is often given insufficient attention. Normally, the volume of waste water to be disposed of equals 50% of the volume of fresh service water used. In the case of the iron and steelworks cited in Table 10 (annual capacity 2.4 million tons of crude steel), this would mean a monthly waste water rate of about 200,000 m³, or 6,500 m³ every 24 hours.

Whereas a certain volume of the total quantity of waste water contains only minor mechanical contaminants, the waste water from the coking plant contains phenol and similar substances; furthermore, oily water, sewage water, and waste water from the blast furnace gas cleaning plant and the generator plant must be dealt with. It will readily be appreciated that waste water of this type cannot simply be discharged into the nearest river—a common practice at the turn of the century.

Numerous efficient waste water cleaning processes can be applied today, and there should be no difficulty in establishing the best process for any given set of circumstances.

In contrast, the *volume* of waste water to be treated may from case to case demand the adoption of special measures.

In the isolated instances where the *full volume* of service water can be taken from a river (no re-cooling), the greater part of the used cooling and service water can safely be discharged into the river again after filtering. Waste water which contains chemicals must first be treated to remove or neutralise acids, phenol, and like substances.

c) Foundry

As a rule, foundries do not constitute an absolutely necessary part of iron and steelworks; some works maintain "repair foundries", i.e. for the production of spare and replacement castings.

Some big iron and steelworks also cast their own *ingot moulds*, but this calls for special casting techniques and equipment, and, above all, experienced personnel. Where these conditions cannot be met, it is much more economical to purchase the required moulds. This applies equally to *cast steel ingot moulds*, which cannot be cast "on the side" using OH steel.

Roll casting is an even more difficult proposition; it is a well known fact that the casting of rolling mill rolls is an art in itself, and demands a high degree of skill and special equipment. If roll casting is to be successful, a complete roll-casting shop must be established and staffed with first-class, experienced personnel. Even then, a longish "warming up" period will be unavoidable, and initial difficulties will cause considerable expenditure. Special rolls—chilled rolls, rolls for hot strip mills etc.—can only be safely tackled after years of experience and numerous trials. It is therefore not surprising that iron and steelworks hardly ever run their own roll casting shops.

The decision as to whether or not the foundry should be operated as a special department for the production of *jobbing castings* must be taken during the main planning stage. Under normal circumstances, iron and steelworks do not run foundries as part of the works itself. As experience has shown, foundries are best operated as independent works, or in conjunction with mechanical engineering concerns.

Metallurgical works with integrated, high-capacity foundries come under a heading of their own. The blast-furnace plant and the foundry are closely linked; the molten iron is transferred to the foundry—often via a hot metal mixer—for casting or for mixing with (hot) cupola furnace metal and then casting. Mixing is effected in the ladle, in flat-hearth type mixers, or in barrel-type hot-metal mixers.

Works of this type usually turn out a high proportion of heavy and medium-heavy mass-produced castings, including spun-cast tubes, water pipes and elbows, branch pipes, and similar articles.

Radiators, for example, are normally excluded from the manufacturing programmes of such works, as the iron analysis must be extremely accurate and the casting temperature very high.

Most of these works produce foundry pig iron only, and therefore have no rolling mills or steelworks. An interesting recent development in this field is the shaking (or centrifugal) ladle for desulphurising and carburising molten iron using lime, coke breeze, soda, etc.

d) Lime Kilns

The lime needed for steelworks and blast furnaces[1] is often supplied in the calcined state, and the iron and steelworks need not instal their own lime kilns. However, if only raw lime is available, suitable lime kilns must be provided for.

If blast-furnace gas, coke-oven gas, or natural gas is available, gas-fired kilns should be given preference to the older coke-fired kilns. This offers certain decided advantages, e.g. calcining is more uniform, no (fuel) ash need be removed, and the lime is not contaminated with sulphur (from coke).

e) Slag Utilisation

The profitable utilisation of slags (blast-furnace slag in the main) can have a decided effect on production costs. Lean ores (Minette, Dogger, and Salzgitter ores, taconite, etc.) naturally produce much more slag than do rich ores, and a good slag economy consequently has an even greater effect in works where this type of ore is smelted.

The nature of the ore and the smelting process applied determine the chemical analysis and physical properties of the slag, and thus its usefulness. The table lists the most common slag products.

Product	Basic slag	Acid slag
Slag granulate sand	×	×
Foamed slag	×	—
Steam granulated slag sand	×	—
Slag sand bricks	×	×
Crushed stone, small stone	×	×
Cement	×	—
Breeze concrete blocks	×	—
Road stone	×	×

[1] Calcined lime is not normally used here; if used at all, it is not fully calcined.

With acid slags, a $CaO:SiO_2$ ratio (p-value) of $p \geqq 0.75$ should generally be adhered to; below this limit, the slag is normally so vitreous that it cannot be used even as crushed (or small) stone.

The special requirements of the various slag treatment processes make a separate slag works necessary—away from the iron and steelworks—and it is also usual to put this secondary works under its own management (see also page 50).

Slag from open-hearth furnaces is always returned to the process or sold as OH slag; consequently, no special treatment is required.

Slag from basic Bessemer converters is normally crushed at the works. The sales price is governed by the content of phosphoric acid. The higher the P content, the higher the sales price.

f) Repair and Maintenance Shop

As the activities of this shop extend to cover *all* departments, buildings, and rooms etc., it may in fact be something more than an ancillary or auxiliary department; in any event, its position in this book should not be taken as an indication of its significance. On the contrary; the work of this department, that of keeping entire works in order, qualifies it for a place among the most important of all departments. However, this book is concerned mainly with the *planning* of iron and steelworks, and not with their operation as such.

The *repair shops* form the backbone of the maintenance organisation. They include shops, personnel and maintenance gangs, for repairing—and to a limited extent for the manufacturing of parts for—structural elements, machinery, crane plant, electric motors and appliances, telephone and measuring systems, woodwork, and so on. A decision should be taken during the planning stage whether one central workshop will suffice, or whether—and this is often made necessary by the distances involved—the blast-furnace plant, for example, ought to have its own fitting shop, electricians shop etc. It must at this juncture be emphasized that *central control of the repair and maintenance organisation is imperative to the smooth running of the iron and steelworks*; this has been clearly proved by experience. In other words, the men employed in the blast furnace department, the steelworks, and so on can be provided with facilities such as workbenches and tool cupboards, but *no independent workshops* should be established. In the interest of a uniform and *economical* repair and maintenance organisation, any disadvantages caused by distances between the various departments must be

9*

accepted with good grace. The establishment of *independent* workshops in each department would be the thin end of the wedge;
sooner or later, the repair and maintenance organisation would be
split up to produce as many smaller organisations as there are
departments.

Spare parts stores also come under the control of the repair and
maintenance department, and should be attached to the central
workshop.

In addition to servicing all crane plant in the works, the repair
and maintenance organisation is often responsible for the crane
operators, too. This is expedient in many smaller works, but not
always to be recommended for bigger establishments.

Nowadays, the *heat economy* or *energy* department comes under the
direct control of the technical head of the iron and steelworks, and
is not supervised by the head of the maintenance section, the usual
arrangement during the early days of development in this particular
field.

g) Transport

1. Outside Transport

Road, rail, and water transport can be used for supplying raw
materials—coal, coke, ore, limestone, dolomite, refractory bricks,
etc.—to the iron and steelworks. Road transport is limited to a
small number of high-grade materials. The choice between rail and
water transport for the remaining bulk goods must be given careful
attention. Where winters are severe, canals and rivers may freeze
up for several months, and the railway must be able to cope with
the extra load during this period; adequate storage facilities must
also be provided to cut the additional load on the railway to a minimum and limit it to unusually long freeze-ups. Calcined lime must
be conveyed in special covered trucks; in countries with severe
winters, thawing bays must be built for the ore and coal wagons.

The transport used for *incoming raw materials* should always be
used for *outgoing products*, e.g. pig iron, blast-furnace slag, crude
steel, basic slag fertiliser, rolled steel, cement, coking plant by-
products. Every effort must be made to ensure full use of the
transport in both directions, a practice that normally results in more
favourable freight terms. High-grade rolled products such as tinplate,
sheet steel for the automotive industry, and special and alloy steels
are nowadays usually delivered "from door to door" by road transport.

2. Internal Transport

As mentioned earlier, careful planning of the transport of *all* materials *to* and *between* the various departments is essential to the smooth running of the works. *Rail transport* is still given preference within the works; cranes also play an important role. Present-day transport and handling facilities in iron and steelworks could in many ways be improved, and a thorough check of existing systems would definitely be worth while.

It has in several cases been proved that *no cranes* are needed in rolling mill finishing shops, for example, except for loading finished products into wagons. Repeated lifting of the rolled material in most finishing departments—two or three lifts by different cranes being the rule—is often excused on the grounds that stocks have to be kept here and there; this argument seldom stands up to close scrutiny. The real reason is to be found in the inexpedient arrangement of the various machines.

Greater use ought to be made of conveyor belts, too; during the past few years, a wide range of conveyor belts has been developed to cater for a multitude of applications.

The *works locomotives* should be under the direct control of the rail transport office, but the drivers must also follow instructions given by the departmental head in whose area they happen to be working; this appears to be natural enough, but is often not put into practice. Many works still calculate their transport costs on the ton-kilometre basis, an absolutely impossible method for *internal* transport; the only practical basis is the locomotive hour. Many iron and steelworks now use container transport systems.

Works harbours must, of course, feature the most up-to-date unloading equipment and bunkers, and should preferably be furnished with conveyor belt systems for the onward conveyance of bulk goods.

Works roadways cannot be too wide. Flyovers should be used at convenient points to keep traffic and works transport apart and allow the workers to reach their places of work on bicycles, and in cars and buses.

Personnel movement should be facilitated not only by providing sufficient roads within the works, but also by establishing adequate transportation facilities between the works and housing estates. Main roads, bus services, tramways, cycle paths and so on cannot be planned too early.

In fact, the *first measure* to be adopted when planning any large-scale industrial undertaking is the building of roads, etc. *to* and *within* the works. This so facilitates the building of the works as such that it always pays to wait until the roads are finished before making the first excavation for foundations, etc.

h) Instrumentation

There are two main reasons for keeping a running check on all operations within the works:

1. The operating crews need *measuring instruments* of all types if they are to run the furnaces, machines, etc. correctly; they also need *regulators* for the automatic correction of gas-air mixtures in gas-fired furnaces, for maintaining the specified furnace temperatures, annealing curves, etc.

2. In addition to controlling operations, modern measuring and regulating equipment can be used for recording and evaluating *cost data*. The data collected by the measuring instruments are transferred direct to counting mechanisms, graphs, and punched cards, etc. This replaces the old system of handwritten records, and provides the management with a complete picture of fuel, power, and material consumption, etc.

The standardisation of measuring and regulating equipment is covered in Section j), page 136.

i) Material Supplies

Adequate supplies of all auxiliary materials required by the works must be guaranteed. It may in some cases be advisable to buy up the supplying works; on the other hand, long-term supply contracts may be better. The most important thing is that a firm relationship be established between the iron and steelworks and the supplying works.

1. Limestone Quarries

Considerable quantities of limestone are needed for the blast furnaces, basic Bessemer, oxygen, and OH steelworks. We take as an example an iron and steelworks with an annual capacity of 1 million tons of pig iron, 0.75 million tons of basic Bessemer and 0.25 million tons of OH (or oxygen) steel p.a. The limestone requirement in this case would be 1,700 tons per day, or 830 kg per ton of crude steel.

The question as to whether the limestone should be calcined (at least for the steelworks) at the quarry or in the iron and steelworks itself must be clarified from case to case. Calcining in the iron and steelworks offers certain advantages, i.e. the calcined limestone is fresh, not having been transported from quarry to works in the calcined state, losses are lower, and as the works-owned kilns are gas-fired, the calcined limestone has a higher degree of purity and does not contain the ash and sulphur contaminations which hand-fired coke kilns always produce. The biggest disadvantage is that the amount of raw limestone to be transported from the quarry to the works is roughly twice that of calcined limestone.

Extra kilns are often needed for calcining large amounts of limestone at the quarry. Normally, they will be hand-fired, as gas will seldom be available at a reasonable price. Apart from the higher transportation costs for raw limestone as opposed to calcined limestone, it is therefore always advisable to calcine the limestone in the works.

2. Refractory Bricks

Adequate supplies of refractory bricks of various grades must also be guaranteed.

Normally, it is not advisable to establish a works-owned factory for the manufacture of these items.

Naturally, the furnaces should be designed or selected to permit the extensive use *standard* of refractory brick sizes and grades.

3. Building Materials

Gravel must be readily available in sufficient quantity to meet the demand not only the initial building stage, but also of numerous building projects in the housing estates, the roads, and the various other industrial works that will be established in the years to come.

The same applies to cement, tiles, bricks, masons' lime, and the many other indispensable building materials. *Bricks* and *tiles* are continually needed in fairly large quantities; consequently, the establishment of a firm relationship with nearby brickyards and tilemakers is particularly important.

Stone quarries must also be given proper consideration.

Plans for obtaining cement, sand, bricks, crushed stone, small stone and so on must be coordinated with the *manufacture of building materials from blast-furnace slag* (see page 130).

4. Foundries

If it is not intended that rolls, moulds, and castings be made in the works, long-term supply contracts should be concluded with reputable foundries for these items, too.

Where high-grade articles such as rolls and moulds are concerned, *quality* must be put before *cost*. The engineers in the various departments must cooperate closely with the purchasing department when normal spare parts are to be ordered, to ensure that orders are placed in accordance with actual demands.

j) Standardisation

All countries have their own *industrial standards*, and naturally enough they must be observed when planning new industrial undertakings. Many major concerns have their own "works standards", which in fact are based on national industrial standards and are therefore nothing more than a concentration of these.

Works that have never established a system of standards should do so when extending or converting the old plant; the system can be gradually introduced into the older parts of the works.

Many appliances, machines, and installations cannot be standardised in the accepted sense and in these cases the equipment used must be restricted to certain *types*. In explanation of what is meant, we give below an example that has long been overtaken by developments, but is nevertheless very instructive.

In 1932, standard widths, perforations, and feed rates were introduced in Germany for the paper used in self-recording measuring instruments. In spite of this, up to 1939 no two firms produced instruments that could use the same paper. The same applied to many other features of such appliances.

The result was that when new iron and steelworks were built, certain *types* of equipment had to be used in order to simplify spare parts stocking, repairs, and permit an adequate degree of exchangeability. Two possibilities were open; each individual department coking plant, blast-furnace plant, steelworks etc., could have been equipped with appliances *supplied by one particular firm*, or *one piece of equipment* could have been obtained for *each department in common* from a number of different firms. The first solution was the more difficult, as the various firms had different production programmes.

In the second case, spares stocking for the entire works was very difficult and the repair of the appliances more complicated.

Generally, the second possibility was therefore chosen, and the entire works equipped with:

Six-colour temperature recorders from supplier "A"
Two and three-colour recorders from supplier "B"
Thermocouple elements, etc. from supplier "C"
Ring balances (volume measurements) from supplier "D"
Ring balances (pressure measurements) from supplier "E", etc.

Naturally enough, not everyone was in agreement with this system, least of all the suppliers, as each firm—quite understandably—wished to sell not only one product, but as much of its entire programme as possible.

During the 25—30 years that have elapsed since then, conditions have radically changed. Nowadays, works can in most cases be equipped with the standardised products of *one* firm, which is undoubtedly the most expedient step to take. As measuring instruments of different makes can now be coupled, e.g. as the value measured by one instrument can be transmitted to a remote indicator of a different make without any difficulty whatsoever, there is no obstacle to the establishment of a *central control point* for evaluating all major measurements from all departments.

Steam pressures, machine sizes, and so on are also standardised as far as possible, and electric motors of certain types and ratings are obtained from the one supplier. Power supplies inside the works are also standardised, e.g. the ring mains at 30,000 V, sub-distribution system 6,000 V, and smaller drive motors are supplied with 500 or 380/220 V. Lighting circuits normally carry 220 V.

The building side can also be simplified. For example, the use of *one crane runway width* (plus two different widths for special cranes) proved to be very advantageous in the case of two German iron and steelworks. They were able to exchange cranes after being damaged during air raids and dismantling had been carried out, and also saved considerable sums by being in a position to order a large number of standardised cranes.

Standardisation should also be effected in cases where an immediate advantage is not apparent; sooner or later, the usefulness of this measure will be demonstrated.

k) Other Facilities, Institutions, and Organisations
1. General

This heading covers such things as:

Works police organisation, works fencing, etc., the works fire service, first aid facilities, possibly a works hospital, and social institutions such as canteens, childrens' nurseries, etc.

These facilities and organisations must be given full consideration at the very outset of the planning stage, in order that the necessary building work can be allowed for. This includes the fence around the works; often, fencing is not readily recognisable as being an important item, but closer scrutiny reveals that it is in fact indispensable. A sufficient number of gateways must be planned, together with suitable connections to existing roads, railways, and so on. Finally, *ample* accommodation must be provided for first aid rooms, the works fire brigade, canteens, and similar institutions; cramped accommodation can be disastrous here. The question as to whether a works-run hospital should be built or whether local hospitals can be used must be clarified on the spot.

2. Supply Organisations

The supply organisations discussed in this section are not connected with the technical side of the iron and steelworks, but demand equal status with technical supply organisations during the planning and building stage.

It will readily be appreciated that the settlement of large numbers of people in areas that previously had no or very few industrial establishments will have to be assisted by the local authorities and the works planning department, at least during the initial period.

In the main, this concerns the provision of food, clothing, consumer goods, and similar articles; after a certain time, private shops, cooperative stores and so on will establish themselves in the housing estates. It is usually not possible to say from the very outset just how many of these and similar services will have to be performed by the iron and steelworks organisation on a permanent basis.

In any case, the initial planning must cater for the accommodation, feeding, clothing, etc. of the building workers and the works staff and their families. Among other things, slaughterhouses, bakeries, cold storage rooms, storehouses, kitchens, and shops have to be built and run. Housing estate plans must also provide for the usual number of shops, etc. per thousand inhabitants.

3. Property Administration

All large-scale projects necessitate the securing of the right of disposal over the land required not only at that moment, but also during the next few decades. This is particularly important in areas where the new works could possibly inconvenience the old-established inhabitants—even though the inconvenience may only be of a temporary nature—and lead to claims for damages being made, with all the usual attendant unpleasantness. Primarily, this concerns damage occasioned by mining subsidence above coal and ore mines. Now mining subsidence can never be fully prevented, no matter how skilfully the mining excavations are planned and executed; the results are often quite serious, of course, especially where buildings are concerned. If property belonging to third parties is damaged in this way, the works may have to pay very substantial damages. In addition, the building and operating of such a gigantic undertaking as a modern iron and steelworks can give rise to considerable annoyances for the local inhabitants. One only need think of obnoxious smells, of dust, noise, waste water and so on, which to a certain extent can also cause damage to property and possibly affect the health of the population.

By purchasing sufficient land around the works, much unpleasantness can be avoided and considerable sums of money saved.

During the first few years after the commissioning of the new works, much of this land will not be needed for building purposes, and other tracts, e.g. those subject to mining subsidence, may never be used for buildings. All such tracts of land must be put under the control of a "property administration office"—the name of the department is not so important—and put to the best agricultural use possible. Experienced farmers should be employed for this purpose; the produce can be sold to the works staff, for example.

4. Housing Estates

As intimated in Section A. IV (Labour market conditions), page 10, the question of housing workers is particularly important in hitherto non-industrial areas. In some cases, it may prove possible to enrol unskilled workers from the local population, and housing need then be provided only for skilled men from other areas.

The housing estates should not be established too far away from the works, otherwise the distance to be travelled daily may be excessive. Neither should they be located on the "leeward" side of the works, where they would be exposed to dust from the converters, etc. Finally, they must be so situated that they blend well into the existing network road and rail communications.

The design of the houses must be left to experienced architects and builders, and should conform to the standard of the country. The housing estates as such must also correspond to the traditional pattern of the country or area in question, and permit rapid settlement of the new inhabitants, who may well come from different areas. For example, the Ruhr miner prefers to live in a small house with a garden and space for keeping rabbits, pigeons, etc., whereas the Ruhr steelworker prefers to live in a flat.

Finally, the housing estates must naturally be provided with modern gas, water, and electric power systems, and the tariff rates for these utilities should be reasonable; schools and similar institutions must be planned in conjunction with the local authorities.

C. The Planning of Iron and Steelworks

As repeatedly stated in the earlier sections of this book, the building of iron and steelworks must be preceded by thorough planning if investments are to bear fruit and the selected technical processes give full satisfaction.

For this reason, the tasks which must be undertaken before the actual investment stage are described below. Naturally enough, when smaller *sections* of an integrated iron and steelworks are erected, e.g. a steelworks using scrap, and a single rolling mill, the *process* side is so clear that significant parts of this preliminary work can be left out. However, the basic conditions, particularly those of a *commercial* nature, must be clarified in advance to ensure that investments are made correctly.

I. Feasibility Study

The first and most important step to take when planning iron and steelworks is the preparation of a feasibility study. The importance of such a study in areas where conditions differ from those encountered in countries with old-established industries—e.g. lack of coking coal, difficult transport conditions—cannot be over-emphasised.

The object of the feasibility study is the fundamental clarification of certain basic factors which govern the chances for success of the project under consideration. It covers the following main points:

a) The domestic market requirement and export possibilities.
b) Availability of indigenous raw and auxiliary materials, or import possibilities.
c) Capacity of the works, composition of the production programme,
d) Processes to be used (ore reduction, steel manufacture, further processing); investigations may also have to be carried out (laboratory, semi-industrial).
e) Rough estimate of investment costs.
f) Rough estimate of production costs.
g) Rough estimate of profitability.
h) Resumé, recommendations, possibly alternative proposals.

It is obvious that the persons conducting studies of this nature must be specialists in the various fields under investigation. Non-industrialised countries and prospective founders of iron and steel-works who lack experience in this particular branch of industry are urgently advised to commission reputable consultants or consulting firms with the preparation of the feasibility study. Iron and steelworks projects are financed with loans, and the banks themselves insist on feasibility studies being made.

Even when loans are obtained from private persons, feasibility studies will be called for.

International banking houses (e.g. the World Bank) invariably reserve the right to approve the firm of consultant entrusted with the preparation of the feasibility study. If the study is not prepared by a neutral firm of consultants—recognised as such in the World Bank classification—the usual practice when the study is prepared by a reputable firm of industrial plant suppliers is that they are paid for the work but do not participate in the actual supply or erection of the plant.

The points listed above under a) to h) are of such importance that they merit closer attention, and for this reason are discussed in detail below.

a) Market Requirement

It must be stressed that the import statistics of many countries often convey an imperfect picture of actual requirements, e.g. the statistics may not provide a reliable guide for the planning of rolling mills. Rolled products are usually grouped in statistical data without relation to the mills in which they are rolled (see section on rolling mills, page 81), or are in part covered in the import figures for finished products. Determination of the actual market requirement for the purposes of selecting suitable steel finishing processes (rolling, drawing, forging, etc.) must be effected with maximum care.

The fullest use should be made of analytical statistics, and the production increase to be expected in the various production branches resulting from the development quota must be given due consideration.

Now the establishment of an iron and steelworks does *not* mean that every type of rolled product required in the country can be manufactured. This would call for a great variety of rolling mills, and it would be quite impossible to fully utilise their true capacities.

The best procedure is to select rolling mills that will cover the domestic requirement for certain items, *and* produce a surplus tonnage for export; the foreign exchange so earned is then used for the importation of the rolled products that cannot be produced economically at home. This naturally calls for a thorough investigation of the import requirements of neighbouring countries. Of course, transportation costs play an important role in this connection, and efforts should be concentrated on countries with which such trading can be carried on at favourable freight rates.

Mention is also made of the fact that neighbouring countries can jointly establish integrated iron and steelworks (with larger capacities) to meet the same ends.

b) Raw and Auxiliary Materials

The properties of the available raw materials must be examined in detail. For example, the bare knowledge that ore deposits are there to be exploited is simply not enough to go on; the full facts must be known. Investigations must be undertaken to establish the magnitude of the deposits, the richness of the ores, and their chemical and physical composition. Surface finds can lead to wrong conclusions; representative samples must be taken. The average values established must be such that the required chemical and physical properties can be attained in the metallurgical process.

All this work, and particularly the taking of samples, must be entrusted to specialists; the analyses must be left to institutes specialising in this type of work.

As it is most unlikely that all the required raw materials such as ores, limestone, dolomite, and coal will be found in any one country, the most favourable sources for the necessary imports must be sought.

Survey work must include both *geological* and *mining* investigations. Plans must be made for the actual mining procedure to be adopted, in order that all cost factors—including the transportation of ores, etc. to the iron and steelworks—may be accurately established.

c) Works Capacity, Production Programme

If no limitations are imposed by the available raw materials, the number and type of rolling mills that will ensure profitable operations can be established, taking into account the domestic demand, exports, and imports; the next step is to plan the capacity of the iron and steelworks to match the rolling mill capacity. Of course, the amount of investment capital available is a major factor, especially when works of this type are established by private persons. However, the expected profits must also be considered when making this decision.

One very important step is the fixing of the *expansion stages*. The initial capacity and the final expansion stage capacity must have a reasonable ratio. For example, an initial capacity of 30,000 t/y and a final stage capacity of 5 million t/y cannot be recommended. A ratio of 1 to 5 for the initial and final stages should not be exceeded under normal circumstances.

The envisaged final stage capacity naturally affects the initial stage, e.g. the *overall size* of the works, and consequently the internal transport distances. Numerous examples could be quoted where the planned final stage capacity was excessive, with the result that operations during the initial stage were uneconomical for many years.

This does not affect the *anticipation* of future *production stages*, e.g. the operation of a rolling mill using imported billets, the later addition of a steelworks using scrap, followed by the erection of reduction equipment.

d) Processes

The envisaged capacity and the available raw materials govern the choice of an ore reduction process, and the steel manufacturing process is then selected to accord with the latter. If the ore and coal analyses deviate from the normal, the classical reduction processes will in all likelihood not be suitable (see also the sections on direct reduction processes, prereduction processes, etc.). In all such cases, laboratory or semi-industrial tests should be carried out to establish whether or not the envisaged processes can in principle be successfully applied, and to obtain technical data such as power consumption in kWh/t for electric processes.

Large-scale industrial tests are also to be recommended when conventional processes are selected, e.g. if ores rich in silicon are to be smelted to the acid process.

Maximum care must be exercised in the selection of technical processes, and this must also be left to experienced consultants.

Expansion stages within the departments should also be carefully planned. We take here the example of converter sizes. Should one operating converter (plus 1 reserve) be installed to match the initial capacity, and a further converter be added during the next expansion stage, or is it more economical to use a bigger converter from the very outset, i.e. a converter that equals the capacity of the second stage?

This would mean having a lower degree of utilisation in the initial stage, but leave a greater margin for a gradual build-up to the desired capacity; also, the investment costs would be higher in the initial stage, but lower in the expansion stage; operating costs would be similarly affected. The time sequence of the expansion stages plays a role in this, too. The linings of converters and mixers can also be thicker in the initial stage to produce favourable metallurgical conditions, but this is an exceptional measure. Finally, mention is made of the fact that the decision as to whether a rolling mill or a continuous casting plant should be used for the production of initial material should be settled at this stage of the feasibility study.

e) Investment costs

At this juncture, it is relatively easy to roughly estimate the required investment capital; naturally, the expected freight costs for the equipment, and the building and erection costs must be included.

f) Production costs

Using the data amassed up to this stage, it is possible to estimate the required labour force; the next step is to calculate the pure production costs.

Normal conditions should be assumed for the feasibility study, i.e. the yardstick to be used is the situation as obtaining after unavoidable initial difficulties have been overcome.

g) Profitability

Taking into consideration service of capital, the estimated operating capital, depreciation and interest rates, the estimated lowest sales prices are calculated. These prices are then compared with the old domestic sales prices for imported products.

It must not be forgotten that the old sales prices for imported items were composed not only of the import prices cif plus transshipment costs and customs duty, but also distribution costs, i.e. in some measure transport costs, in great measure sales organisation costs.

The new iron and steelworks must either use the same organisation and allow for these costs, or establish its own organisation and include the costs so incurred.

Now even though the profit to be allowed for is a very important item, it must be remembered that iron and steelworks cannot earn profits as easily as the customer goods industry, for example, because of the huge sums involved in depreciation. In many instances, tariff protection is granted by the state (the party most interested in saving foreign exchange) for at least the initial period, when costs will definitely be higher than normal.

In countries with limited industrial experience, foreign specialists, engineers, and other senior staff will be needed for a certain number of years—the exact period is governed by the nature of the process. The costs so incurred can be quite considerable, and must be taken into account.

h) Summary

Study of all data accumulated up to this juncture will permit the conclusion as to whether it would be commercially sound, paying due attention to technical requirements, to erect a works with the envisaged capacity, programme, process, etc., or whether alternative proposals should be made.

It will also now be possible to say whether, when, and to what extent further investigations should be made, e.g. additional technical investigations, the selection of various sites, and test drillings to establish ground data.

A correctly prepared feasibility study will in any event enable the customer, the financing institute, and the state to decide whether the project should be pursued, and whether any modifications to the original plan are called for.

Both the customer and the firm or persons commissioned with the execution of the feasibility study must naturally be prepared for requests for further investigations from financing institutions and government departments, e.g. those concerned with the granting of import licences etc.

Naturally, the question of profitability can be of secondary importance in projects such as these, which are often financed by the state itself or with state aid. For example, a pressing need to save foreign currency, or other political reasons may exist.

II. Preliminary Planning

Once the feasibility study has been given the approval of all concerned, it is advisable to proceed with the preliminary planning stage. This applies to bigger projects; smaller, uncomplicated projects may not need such a stage.

During the preliminary planning stage, numerous assumptions made in the feasibility report are resolved by making further investigations, and various fundamental matters such as plant siting are clarified. When these tasks are completed, a document is drawn up which, after being approved by all concerned, serves as the foundation for the *detailed* and *final* planning of the project. *Final planning* incurs a tremendous amount of work, particularly in connection with auxiliary and ancillary plant and equipment; consequently *every detail* must be meticulously examined during the *preliminary* planning stage. It may even be necessary to carry out *large-scale tests* to prove the suitability of the envisaged technical processes, for example.

In the feasibility study, the plant and equipment is briefly described, but in the preliminary planning stage, rather detailed specifications must be prepared. In the case of smaller plants, these documents are often used as tender specifications. When all open questions have been clarified, tenders can at least be put out for cost estimates (larger items of equipment) in order that the estimated investment costs may be confirmed.

III. Final Planning

As soon as the preliminary planning stage has been completed, i.e. when all technical matters, including site and ground suitability have been satisfactorily settled, *final planning* is undertaken; the *finished project* is given its shape.

Depending on how the order is to be placed, the final project document is used for ordering from *one firm*, or for the appointing of *one firm as main contractor* for the *entire* iron and steelworks. This is possible where the ordering party has a staff of experienced engineers at his disposal who are capable of conducting all negotiations before and during the erection up to commissioning. If this is so, the final planning need not be very detailed.

On the other hand, if it is intended that a world-wide tender be published, the orders for various items of equipment going to different firms in a number of countries, the final planning must be

10*

carried out in very great detail. For example, the prospective
suppliers must be put in a position to submit quotations in such
detail that the individual items of plant and equipment (main plant,
e.g. ore reduction facilities, steelworks, rolling mills, and auxiliary
equipment, — for power generation and distribution, water supplies, waste water disposal, gas distribution, laboratories) can be
so supplied and erected that the completed iron and steelworks
will function correctly and in harmony.

The appointing of a consulting firm to assist the future works
management is a common practice. The consulting firm functions
more or less as the management, checks all quotations, makes proposals as regards ordering, carries out final acceptance inspections,
supervises the erection and building work, organises the commissioning, and sometimes makes arrangements for a management for
the first few years of operation. During this period, the actual
owners act as the senior supervising body only. Even though this
is a very expensive procedure, the appointing of consulting firms
of this type to look after the interests of the owners is to be highly
recommended; it ensures that these intricate projects are carried
out smoothly and expertly.

Final planning covers the following tasks:

All specifications for main, auxiliary and ancillary plant and
equipment.

Layout plans of the entire works, with typical drawings of the
individual items of equipment.

Flow diagrams for all materials, including home scrap, etc.

Outline foundation drawings with load data, details of any special
conditions (heat, etc.), and ground data; these drawings must be so
complete that efficient building contractors can submit quotations
for the planning stage (shuttering and reinforcing plans, static calculations, etc.) and carry out the actual work.

Schedules with details of despatch, erection, building and other
deadlines up to actual commissioning.

Programmes for the training of personnel in the countries (or
works) supplying the equipment.

Complete staffing schedules (including those for outside personnel).

Listing of all items to be obtained locally for erection purposes,
e.g. electric power, water, derricks, erection tackle, provisional
storerooms and storage yards.

Coordination plans for such things as standards (uniform pipe-work and fittings, valves, pumps, couplings, etc.).

Drawing up quotation submission details, e.g. payment conditions, penalty clauses, bank guarantees, etc.

Drawing up details of guarantees for the equipment as such, together with performance guarantees, etc.

Drawing up conditions for the provision of personnel, e.g. erection staff, commissioning staff.

Details of references, etc. to be provided.

These are the main points. Naturally, if a turn-key plant is ordered from *one concern*, quotation data and indeed the final planning work as such can be simplified. However, as the supplier of a turn-key plant has to bear "global" responsibility for the project, his terms are normally somewhat higher.

Where very large works are concerned, it is not usual to place the order with a single firm, but a consortium of firms. The consortium is often composed of firms from different countries (credit reasons). In such cases, it is most important that one firm be appointed as leading and coordinating member—carrying the main responsibility—and that the spheres of responsibility of the other firms are carefully coordinated; this task can be entrusted to a consulting firm. When big integrated iron and steelworks are being built, several hundred people are often present on the site; naturally, this calls for excellent organisation and coordination, particularly when several nationalities are involved.

IV. Commissioning

As intimated in the preceding sections, both the supplier and the ordering party must carefully prepare the way for the commissioning of the works.

Not only must certain members of the staff be given training in the supplying works, and operators be engaged from other countries or areas, but some of the erection personnel will have to be retained after erection and commissioning to instruct the staff in maintenance and repair work, and to remedy any initial defects that may occur. All new iron and steelworks should engage an ample number of such specialists and operators on long-term contracts; some will be needed for several years. Many works managements who where of the opinion that this was unnecessary have had to foot heavy bills when the planned annual output was not attained (quantity and quality), and subsequently had to engage specialists on less favourable terms than would have been possible earlier.

Apart from normal guarantees for the quality of the machines etc. as such, the obligations of the supplier usually terminate on the performance tests being successfully completed. However, performance tests are normally limited to the attainment of the design outputs over a limited number of shifts or days, and are carried out under conditions that are exactly determined in advance.

The future works management also has certain specific obligations, e.g. raw materials of the correct quality must be made available in such quantity that operations can proceed smoothly. Now an iron and steelworks is much more sensitive to unforeseen interruptions than are mechanical engineering concerns, for example. One only has to think of the number and nature of the *hot units* (blast furnaces, etc.) to realise the damage that can be caused by stoppages of any length. On the other hand, the management has certain tasks of a commercial nature, e.g. the provision of sufficient operating capital; the amount required depends on such things as purchasing, storing, and sales conditions. For example, an iron and steelworks working very closely with an ore mine need not tie up capital for ore supplies covering a period of several months; but if ores must be transported over long and perhaps difficult routes (winter), or the mines cannot work during very cold weather (open-cast workings), a considerable amount of capital will be needed to ensure the ready availability of adequate stocks.

The same applies to the sales side. If it may be expected that the various products will find a ready market, so ensuring steady returns, no capital is required; but if periods of stagnation in the market are to be feared, then funds must be provided to cover the gaps.

As a rule of thumb, the operating capital can be taken as equalling some 15—20% of the investment costs under normal circumstances. Allowance must naturally be made for the fact that bigger works usually do not attain full production until some months or even years have elapsed; this depends, of course, on how complicated the works is, and on the amount of money previously invested in training and management.

It is obvious that big works should not be commissioned in a single day, but in such stages that effective control is guaranteed.

Departments that mesh together must start up in rhythm unless storage facilities exist or the products can be put to some other use.

For example, the steelworks will have to start up shortly after the blast furnaces have been commissioned unless the blast furnace metal can be put through a pig-casting machine or otherwise profitability used.

The works management must also ensure that the various metallurgical units are heated at the specified time prior to commissioning, and that this work is correctly supervised. This includes such things as drying the blast furnaces, converters and so on, the end of which normally merges with the commencement of actual operations.

It is also vitally important that workshops and cranes are available long before the commissioning date for the works as such. The workshops are needed for the numerous minor modifications that always have to be taken care of during the erection phase, and the cranes are needed during the same period. Consequently, arrangements must be made in good time for the final inspection of such facilities and equipment, and for the timely transfer of ownership.

Finally, sufficient spare parts for at least one year of operation must be available on the works starting up, or drawings must be bought and manufacturing facilities arranged to ensure that the required parts can be made on the spot. This is a very important point, but is nevertheless often overlooked by the customer.

D. Investment Costs

The term "investment costs" is to be understood as embracing all capital expended in the building of the iron and steelworks. Normally, the purchase price for land (site) and the costs of any extensive earthmoving work that may be necessary are left out. In the following, the investment costs cover in the main the following plant and equipment:

a) Coking plant.
b) Ore preparation plant (crushing, screening, sintering, and pelletising plant etc.).
c) Pig iron production plant (ore reduction plant, blast furnaces, electric pig-iron furnaces).
d) Steelworks plant:
   1. OH steelworks.
   2. Basic Bessemer steelworks.
   3. Oxygen steelworks (top-blown oxygen steelworks, LD, LDAC, OLP, Kaldo, etc.).
   4. Electric steelworks.
e) Rolling mills, continuous casting plant.
f) Auxiliary and ancillary plant (see also Section B. III, page 112, and Table 16—page 163).
g) Power station.

In the following, some essential data and rough estimates needed for establishing the investment costs for the above plant and equipment are given.

The costs quoted below are typical examples taken from West European iron and steelworks and date from 1964—65. It must be borne in mind that these costs are *subject to the same changes* as occur in the prices of all goods during the course of time. The prices quoted can also generally apply free Atlantic seaport. For plant shipped overseas, e.g. to Africa, South America, Asia, and Australia, the applicable freight costs must be added. It must also be remembered that low labour costs at the erection site can reduce expenditure.

The total costs of turn-key iron and steelworks erected in Western
Europe can in general be subdivided as follows:

fob supplies	50—70%
erection	15—10%
building work (found- ations, buildings, etc.)	35—20%

I. Coking Plant

Normally, the iron and steelworks planner is not responsible for
coking plant, even though the erection of a works-owned cokery
may appear expedient. The cost of such a plant, including the
usual by-products facilities, can be estimated at roughly 20—25 $
(80—100 DM) per ton capacity (coke). This is based on coal inter-
nationally recognised as "coking coal", which has a yield of about
75—77% (i.e. one ton of coking coal yields some 750—770 kg coke).
Depending on the quality of the coke, some 5—10% of coke breeze
(0—10 mm) must be deducted from gross coke tonnage.

II. Ore Preparation and Beneficiation Plant

This heading includes all equipment for ore preparation, i.e.
crushing and screening to the required lump size, and the sintering
and pelletising equipment used for agglomerating the ore fines.

Beneficiation plant is used for upgrading ore by separating out
undesirable gangue elements and—especially in the case of rich
ores—by prereduction.

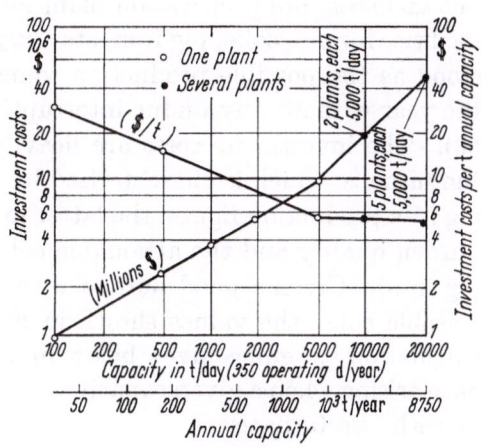

Fig. 29. Investment costs —
total and per ton capacity — of
pelletising, grate-belt, and
grate-kiln plants (see J. ASTIER,
L'agglomération en boulettes
des minerais de fer, Revue de
Métallurgie, May 1962, vol.
5 pp. 417/435).

Investment costs vary within a very large bracket, as the facilities are individually designed for specific types of ore.

The following values can be taken as a guide:

Plant	Related to	Total investment costs per annual ton capacity	
		$/t	DM/t
Ore preparation plant			
Ore beds	Crude ore	2.00	8.—
Crushing and screening plant	Crude ore	5.00	20.—
Sintering plant	Sinter	7.50	30.—
Pelletising plant[a]	Pellets	7.50 to 10.—	30.— to 40.—
Ore beneficiation plant			
Wet magnetic upgrading	Charge	6.25	25.—

[a] see also Fig. 29.

III. Pig Iron Production Plant

a) Ore Reduction

Prereduction plants working in conjunction with electric reduction furnaces have already been discussed. Plants turning out a product rich in Fe—sponge iron—by direct reduction are so few in number that no reliable figures can be quoted at the present time.

b) Blast Furnace Plant

The investment costs for blast furnaces are to be understood as comprising the complete costs of the turn-key plant including all auxiliaries, the hot-blast stoves, blowers, gas cleaning plant etc., but *excluding* ore preparation plant and all transport and handling facilities for ore, coke, pig iron, etc. *Pig-iron capacity* is to be understood as the possible production tonnage working on 360 days in the year without any undue interruptions (see also Fig. 13 on page 53). The investment costs are heavily influenced by the size of the furnace, which is characterised by the hearth diameter and the pig iron production figure, the latter in turn being influenced by the burden quality and the attainable coke rate (in kg coke per ton of pig iron). Given a good burden with a net yield of 54—60%, and suitable coke, the values shown in Fig. 30 can be achieved. The calculation examples given below in Table 11 show the cost difference arising when a given capacity is to be achieved with one furnace or with two furnaces.

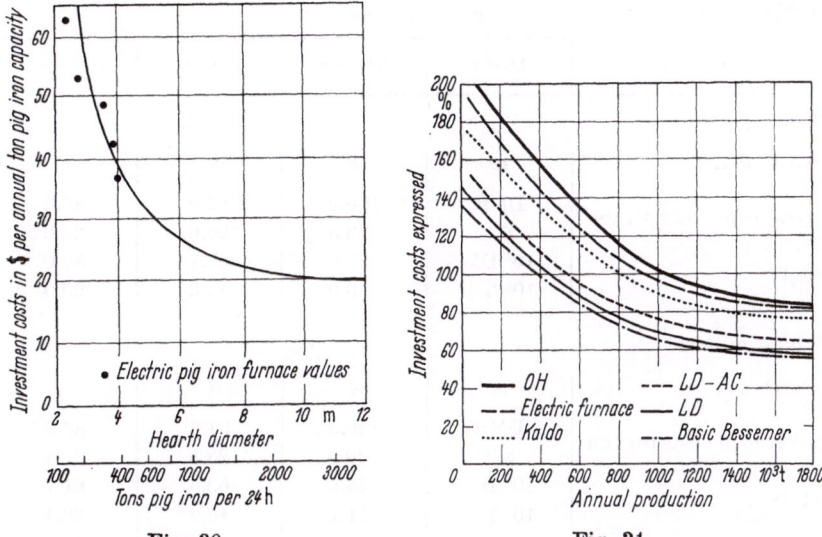

Fig. 30.   Fig. 31.

Fig. 30. Blast furnace investment costs per annual ton pig iron capacity.
The *following investment costs include:* The complete blast furnace with burdening
equipment, gas cleaning equipment, blowers, cowpers, cooling system, furnace gun,
and all electricals.

Excluded: Bins for ore, coal, etc., gasometer, slag granulation, pig casting machine, etc.

Fig. 31. Investment costs for various processes; the investment costs for an OH
steelworks with an annual capacity of 1 million tons are taken as equalling 100%.
(From "Comparison of Steelmaking Processes", published by the Economic
Commission for Europe of the United Nations, 1962.)

Table 11. *Blast furnace investment costs*

Hearth diameter m	Output in t/24 h[a]	Pig iron[b] t/y	DM/t[c]	$/t[c]	Investment costs	
					million DM	million $
3	230	83,000	220.—	55	18.0	4.5
4	400	144,000	156.—	39	22.5	5.6
5	640	230,000	124.—	31	28.5	7.1
6	930	335,000	106.—	26.5	35.5	8.9
7	1,250	450,000	94.—	23.5	42.0	10.5
8	1,650	590,000	86.—	21.5	51.0	12.75
9	2,100	750,000	82.—	20.50	61.5	15.4
10	2,600	930,000	81.—	20.25	75.0	18.75
11	3,100	1,120,000	80.—	20	90.0	22.5
12	3,700	1,330,000	80.—	20	106.0	26.5

[a] Average values based on furnaces with coke rates of 600 kg/t pig iron and
hearth loads of 1,000 kg/m²h (see Fig. 13, p. 53).

[b] Taken from graphs in Fig. 30.

[c] 360 operating days per annum.

(Table 11, cont'd)   *Calculation examples:*

Output in	t/year	300,000	700,000	1,000,000
1 Blast furnace				
Hearth dia.	m	6	9	11
Investment costs per ton	DM/t	106.0	82.0	80.0
	$/t	26.5	20.6	20.0
total	10^6 DM	31.8	57.4	80.0
	10^6 $	8.0	14.3	20.0
2 Blast furnaces				
Hearth dia.	m	4	7	8
Investment costs per ton	DM/t	156.0	94.0	86.0
	$/t	39.0	23.5	21.5
total[a]	10^6 DM	44.5	62.5	81.7
	10^6 $	11.1	15.6	20.4

[a] With two blast furnaces, the investment costs can be cut by about 5%. This is taken into consideration in the total sum (million DM) e.g. $300,000 \times 156 \times 0.95 = 46.8 \times 0.95 = 44.5$ million DM, or 11.1 million $.

IV. Steelworks

Fig. 31 gives an insight into the investment costs as applicable for various steelmaking processes. This chart is an extract from the "Comparison of Steelmaking Processes". As the values are expressed as percentages, one million tons of OH steel p.a. being taken as 100%, the curves have a limited value only. They merely show that the investment costs of OH furnaces are the highest, and that those of converting steel mills are the lowest.

a) Open Hearth Steelworks

Table 12 gives examples of OH steelworks equipped with various numbers of furnaces of different capacities. The values given are based on experience gathered in Europe and on the mode of operation usual in that part of the world, i.e. the pig-and-scrap process. The charge is composed of 300 to 400 kg hot metal (or pig-iron) per ton of crude steel, the balance being scrap. Fig. 18 shows furnace outputs in t/h as a function of the furnace capacity (= tap weight). It may be presumed that the outputs per hour given in this figure will not deviate greatly from European values when using pig iron processes, which are particularly common in the USA. Table 13 shows the investment costs for mixers, OH furnaces, buildings, etc.

Table 12. *Investment costs of OH steelworks with furnaces of various sizes*[a]

| OH furnace capacity | No. of furnaces | Total furnace capacity | Annual capacity[b] | Investment costs per ton | | | |
| | | | | per ton capacity | | total | |
t	Units	t	t/y	DM/t/y	$/t	million DM	million $
100	2	200	200,000	170.—	43	34	8.5
	4	400	400,000	158.—	46	63	16
	6	600	600,000	155.—	39	93	33
200	2	400	360,000	145.—	36	52	13
	4	800	720,000	135.—	34	97	24.5
	6	1,200	1,080,000	132.—	33	142	35.5
300	2	600	430,000	125.—	31	54	13.5
	4	1,200	860,000	116.—	29	100	25
	6	1,800	1,290,000	114.—	28	147	36

[a] These values relate to OH furnaces working on pig-scrap charges with an average of 350 kg pig iron per ton of crude steel.

[b] The capacity of an OH furnace is   14 t/h for 100-ton furnaces
                                     25 t/h for 200-ton furnaces
                                     30 t/h for 300-ton furnaces

These values were obtained in recent operations (Autumn 1965). 300 full operating days per annum (7,200 h/y) were taken as a basis (see Fig. 18).

Table 13. *Breakdown of investment costs for OH steelworks with furnaces of various sizes*

OH furnace capacity	t	100	200	300
Mixer plant	%	2	2	2
OH furnaces + oil firing	%	31	27	25
Buildings + cranes	%	35	35	35
Casting bay + ladle equipment	%	10	12	13
Building work, miscellaneous	%	22	24	25

b) Electric Steelworks

Integrated iron and steelworks are normally equipped with electric arc furnaces with capacities of between 30 and 150 tons. Fig. 32 gives the estimated investment costs *less* the power station section.

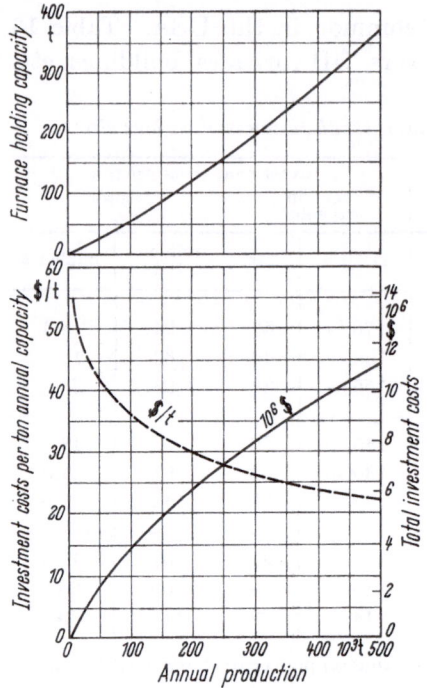

c) Basic Bessemer Steelworks

Even though the basic Bessemer process is being replaced by the oxygen steel process, some data is given below on the former process, in view of the significance it enjoyed in Europe during the past 100 years or so.

Depending on the mode of operation, basic Bessemer steelworks produce about 4—7,000 t crude steel per ton converter

Fig. 32. Electric steelworks. *Above:* Relationship between holding capacity and production capacity; *below:* Investment costs (total and per ton annual capacity).

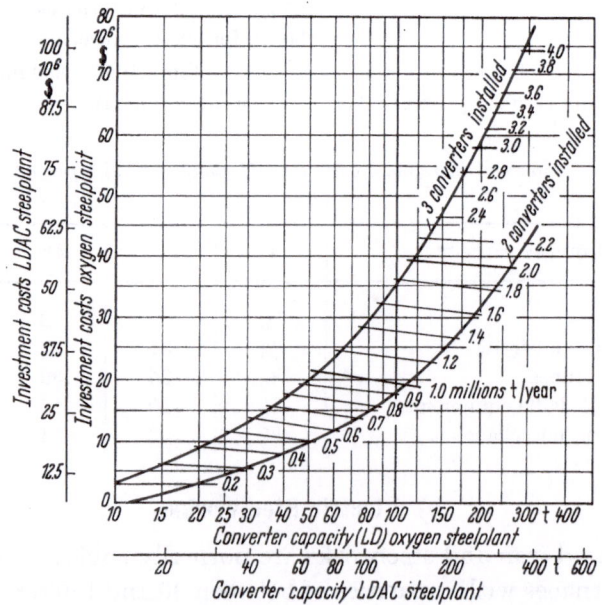

Fig. 33. Investment costs and annual outputs of oxygen (LD) and LDAC steel plants.

capacity per annum. The investment costs lie between about 20 and 25 $ (80—100 DM) per annual ton capacity.

d) Top-blowing Oxygen Steelworks

Fig. 21, page 75 shows the possible capacities, and Table 14 the investment costs (total and per ton capacity) for 2 and 3-converter shops, the converter capacities being 30 to 300 tons.

Fig. 33 shows the investment costs of oxygen steelworks with 2 converters (1 in operation) and 3 converters (2 in operation). The curves show that the costs for the 3-converter shop are higher than those of 2-converter shops, the capacities being the same. As capacity increases, balance is achieved, the costs being equal at about 2.4 million tons.

Table 14. *Investment costs of top-blowing oxygen steelworks*

Capacity per converter	Capacity t crude steel/y[a] (300 days/y)	Investment costs per ton capacity		Total investment costs	
		DM/t	$/t	million DM	million $
2-converter plant (1 in operation at any one time)					
30	310,000	141.—	35.0	44	11.0
50	510,000	118.—	29.50	60	15.0
100	965,000	93.—	23.0	90	22.5
150	1,360,000	88.—	22.0	120	30.0
200	1,710,000	86.—	21.50	147	36.75
250	2,000,000	85.—	21.25	170	42.5
300	2,250,000	82.—	20.50	185	46.25
3-converter plant (2 in operation at any one time)					
30	590,000	125.—	31.25	74	18.5
50	960,000	104.—	26.0	100	25.0
100	1,810,000	88.—	22.0	160	40.0
150	2,500,000	86.—	21.5	215	53.75
200	3,160,000	81.—	20.25	256	64.0
250	3,670,000	80.—	20.0	295	73.75
300	4,100,000	80.—	20.0	330	82.5

[a] A tap-to-tap time of 50 to 65 minutes was taken as the average over the year, with 7,200 operating hours per year (see also Table 5 and Fig. 21). If the converters are available, and the oxygen plant is designed to cope with the demand, 1½ converters can also be used; however, this cannot be regarded as a permanent solution, and is merely mentioned in passing.

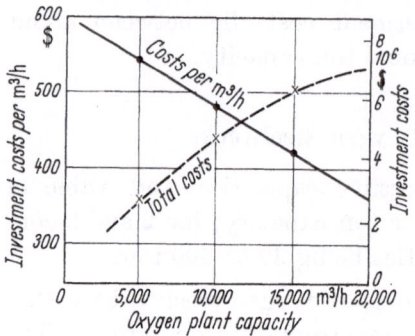

Fig. 34. Investment costs of complete
oxygen plants.

As shown in Fig. 33, the converters in LDAC shops must be about 30% bigger to achieve the same capacity as the LD units; this increases the investment costs by 25—30% when compared with oxygen steelworks.

The following list gives a breakdown of the investment costs of oxygen steelworks; the individual parts are assumed to be finish-erected.

Breakdown of investment costs of an oxygen steelworks

	%
Mixer plant	2
Oxygen converters and charging equipment	10
Dedusting plant	8
General steelworks equipment	11
Electricals	6
Cranes, buildings	35
Oxygen plant	12
Other building work	16

Fig. 34 gives the investment costs oxygen plant for oxygen steelworks.

V. Rolling Mills

Rolling mills are so diverse in type that the quoting of investment costs is extremely difficult. To this is added the difference in cost between small open or semi-continuous mills on the one hand and fully mechanised mills on the other. The overall investment costs for normal rolling mill plant can be broken down as follows:

	%
Mechanical part	approx. 35
Electricals	approx. 20
Other equipment	approx. 10
(furnaces, roll turning lathes, workshops, compressors, water supplies)	
Foundations, buildings, cranes	approx. 35

This breakdown does *not* include mechanised high-performance rolling mills, the mechanical and electrical equipment of which often accounts for two-thirds of the total investment costs.

In contrast to blast furnace and steelworks plant, the relationship between the price and (increasing) capacity of rolling mills cannot be plotted on a graph; as indicated in the section on rolling mills, the capacities of these installations increase by leaps and bounds. It was, however, felt that this book ought to give some information on this point, too, and guide values for the investment costs of the more common rolling mills can be found in Table 15. The capacities listed give an indication of the mill size.

Table 15. *Rolling mill investment costs*

Rolling mill	Capacity 1000 t/y	Investment costs per ton capacity	
		DM/t	$/t
Heavy slabbing mill	2,000	32.—	8.0
Heavy blooming mill	1,000	36.—	9.0
Medium blooming mill	500	44.—	11.0
Light blooming mill	300	52.—	13.0
Heavy plate mill	300	200.—	50.0
Heavy plate mill	1,000	150.—	37.50
Hot wide strip mill	1,000	150.—	37.50
Hot wide strip mill	2,000	120.—	30.0
Steckel mill	300	200.—	50.0
Cold strip mill[a] with 1 reversting and (1,500 mm)	300	500.—	100.0
with tandem mill (1,500 mm)	1,000	360.—	90.0
Open sheet mill[b]	70	300.— to 500.—	75.0 to 125.0
Bloom/billet mills (1-stand)	300	80.— to 100.—	20.0 to 25.0
Bloom/billet mills (3-stand)	300 to 400	100.— to 120.—	25.0 to 30.0
Bar and section mills:			
Heavy mills	300	100.—	25.0
Medium mills (open)	100	200.—	50.0
Medium mills (continuous)	400	150.—	37.50
Light mills		(see Fig. 35)	

[a] Including pickling plant, (hood) annealing plant, and skin-pass stand, but excluding tinning or galvanising lines.

[b] Normally comprising a Lauth 3-high mill with two 2-high stands and a 2-high skin-pass stand.

Fig. 35 shows an approximate relationship between investment costs and increasing capacity for *light section mills*; the degrees of mechanisation naturally overlap. Particularly in the case of section mills (all types), the actual rolling times and consequently the

production rates are heavily influenced by the make-up of the
rolling programme (average weight per metre) and the sizes of the
orders (frequency of roll changing).

Example. A section mill has a rolling programme in the bracket 10 t/h to 30 t/h
(actual rolling time — taking into consideration the number of passes required
and the desired weights per metre). The upper limit is assumed to be set by the
maximum furnace capacity of 20 t/h.

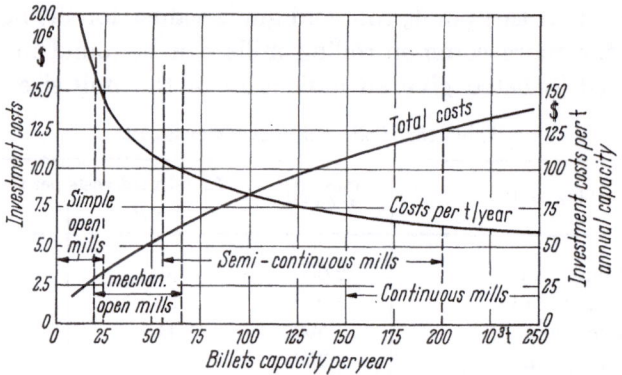

Fig. 35. Roughly estimated investment costs for light section mills.

Assumption a). The programme lies on the upper limit, and the
orders are of such a size that all roll changing work, etc. can be
carried out on Sundays. This gives 6 days at 24 h over 32 weeks
= approx. 7,500 rolling hours p.a. At a utilisation figure of approx.
80% (adjusting passes, minor defects) we obtain 6,000 h/y × 20 t/h
= 120,000 t/y.

Assumption b). The programme lies on the lower limit, and the
orders are such that the rolls, etc. have to be changed rather frequently. As a result, the mill can only be operated on two shifts per day.
This gives 6 days at 16 h over 52 weeks = approx. 5,000 rolling
hours p.a. Using the same degree of utilisation, we obtain 4,000 h/y
× 15 t/h = 60,000 t/y.

The costs of rolls for *section mills* can be extremely heavy if
every section in a very wide programme is to be catered for (important point when installing the first rolling mills in any particular country).

Tube rolling mills are so diverse in type that they cannot be
treated generally.

Hot wide strip mills are of either semi or fully continuous design. The former have lower capacities and cost less, even though the costs per ton capacity are somewhat higher. It is assumed here that the most widespread barrel length of approx. 1,650 mm (66″) —giving a strip width of 1,500 mm—is used.

The greater proportion of coils produced in hot wide strip mills is rolled down to quality sheet with an average gauge of 0.8 mm (with very wide margins above and below this figure) in cold wide strip mills; the remainder is rolled down to 0.24 mm (average) and used for the manufacture of tin plate.

Table 16 lists the costs of these cold rolling mills —a cold wide strip mill (reversing) of low capacity, and a four-stand tandem mill with an annual capacity of 1 million tons.

Table 16. *Investment costs of cold wide strip mills*
a) Four-stand tandem mill; 1,500 mm, 1 million t p.a.;
b) Cold strip reversing stand; 1,500 mm, 250,000 t p.a.

Item	Capacity			
	a) 1 million t			b) 250,000 t
	million DM	million $	%	million $
Cont. pickling plant	17	4.25	5	4.0
Four-stand tandem mill	46	11.50	13	
Reversing stand	—	—	—	3.0
Cleaning line				
Cover annealing plant				
(90 bases)	27	6.75	7	1.5
2-4 skin-pass stands	42	10.50	12	1.7[a]
2-4 sheet dividing lines	15	3.75	4	3.6[a]
2-4 sheet slitting lines	12	3.0	3	3.2[a]
	159	39.75	44	17.0
Bays	100	25.0	28	
Cranes	36	9.0	10	35.0
Foundations	30	7.50	8	
General facilities	35	8.75	10	
Totals:	360	90.0	100	

[a] for one unit only.

The following table gives the estimated investment costs for the galvanising and tinning lines used in conjunction with the 250,000 ton cold rolling mill listed in Table 16.

	%	%
Reversing cold strip mill		100
Processing equipment:		
Galvanising (15,000 t/y)	4	
Electrolytic tinning line (50,000 t/y)	33	
Hot-dip tinning line (10,000 t/y)	4	
Auxiliary plant	9	50
Entire plant		150

VI. Power Station

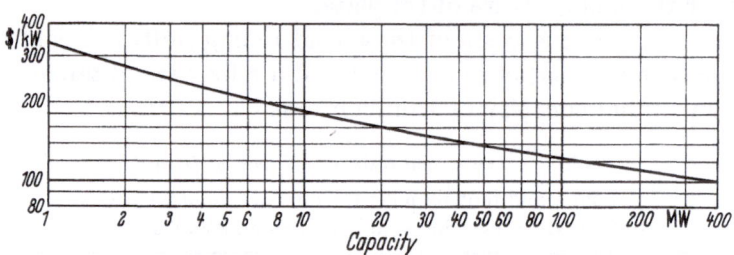

Fig. 36. Investment costs of power stations (by K. Schäff, Steag).

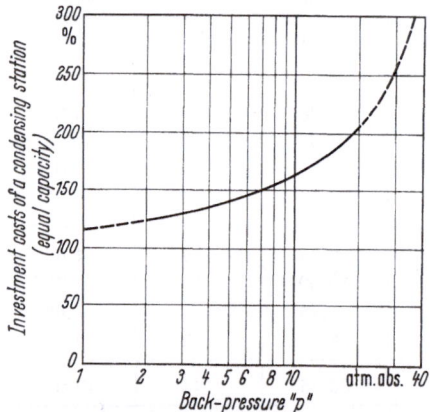

Fig. 37. Investment costs for back-pressure
power stations (by K. Schäff, Steag).
(Condensing power stations with fresh
water cooling = 100%).

The significance of the works power station is discussed in Section B. III. a), page 116, and the reasons why iron and steelworks power stations are comparatively small are also given. Figs. 36 and 37 list the roughly estimated investment costs for power stations of this type (for both condenser and back-pressure plant).

VII. Buildings

The costs of the various buildings are estimated below:

Type	Weight per m² in kg/m²	Costs per m²	
		DM/m²	$/m²
Steelworks buildings	1,300	1,500	375
Heavy rolling mill buildings	1,000	1,200	300
Light buildings	800	1,000	250

The following also applies to rolling mill buildings:

	m² per 1,000 tons capacity m	Costs per ton capacity	
		DM/t	$/t
Blooming mill buildings	11	14	3.50
Continuous billet mill buildings	11	14	3.50
Bar and section mill buildings	75—135	90—165	23—41
Ditto, average	100	120	30

VIII. Auxiliary and Ancillary Departments (General)

These departments account for about one-third of the total investment costs of an integrated iron and steelworks; their costs can be estimated at 75 to 125 $ (300.— to 500.— DM) per ton capacity (crude steel). Table 17 lists the investment costs of the auxiliary and ancillary departments of an integrated iron and steelworks with an annual steel capacity of 2.4 million tons; included in this list is a coking plant to cover the full amount of coke required, and a power station with a capacity of about 200 MW to cover the full power requirement. The table also shows the percentage figure (of total investment costs) for each individual department. For example, the transport and handling facilities and energy distribution systems account for about three-quarters of the investment costs, or 55 $/t steel (220 DM/t). It is here that efforts must be made during the planning stage to keep costs down. The next biggest item is listed under 1): storage facilities, etc. for raw materials such as ore, coking coal, coke, limestone, etc., which account for about 7% of the total investment costs in this sector. The costs for social amenities (13) are roughly the same. As the distances between the various departments of a modern iron and steelworks are quite considerable, the washrooms, canteens, first-aid rooms, etc. must be distributed around the works and allocated by departments.

IX. Total Investment Costs of Integrated Iron and Steelworks

Table 18 lists the investment costs of iron and steelworks with various capacities built in recent times (1963—1965). Not all these works had their own coking plant, but each one incorporated blast-furnace plant and oxygen steelworks. The rolling mill plants differed, i.e. the works quoted under Example 1 rolled bar and sections

Table 17. *Breakdown of investment costs for the general departments and facilities of an integrated iron and steelworks* [a]

		Stage I per t capacity (crude steel)			Final stage per t capacity (crude steel)		
		DM/t	$/t	%	DM/t	$/t	%
1	Raw materials storage, including handling and intermediate storage	30.—	7.50	7.5	20.—	5.0	6.7
2	Steelworks scrapyard	1.—	0.25	0.2	1.—	0.25	0.3
3	Transport and handling facilities (works railways, conveyor belt equipment, works roadways with wheeled vehicles)	130.—	32.50	32.3	100.—	25.0	33.4
4	Laboratories, quality testing department	10.—	2.50	2.5	7.50	1.90	2.5
5	Power and energy distribution (electric power, water, gas, etc.)	155.—	38.75	38.5	120.—	30.0	40.0
6	Main repair workshops, spare parts stores	20.—	5.0	5.0	10.—	2.50	3.3
7	Repair and maintenance department	5.50	1.4	1.4	4.—	1.0	1.3
8	Building dept., joiners' shop	3.—	0.75	0.7	2.—	0.50	0.7
9	Main stores	2.—	0.50	0.5	1.50	0.36	0.5
10	Communications (telephone, pneumatic-tube systems etc.),	2.—	0.50	0.5	1.50	0.37	0.5
11	Fire service, industrial police dept., accident prevention, etc.	1.50	0.35	0.4	1.50	0.37	0.5
12	Administration, guest house, etc.	12.—	3.0	3.0	6.—	1.50	2.0
13	Social amenities (washrooms, canteens, first-aid facilities)	30.—	7.5	7.5	25.—	6.25	8.3
Total:		402.—	100.50	100	300.—	75.0	100

[a] In the first stage, this works has a capacity of 800,000 t/y; at this point, various plant and equipment must be installed in anticipation of the final expansion stage (2.4 million t/y).

only, whereas the other two works turned out flat products only; one of these (Example 3), is also heavily engaged with the processing of cold rolled strip. Furthermore, the power station of the smallest works was designed to cover only a fraction of the demand, whereas in the other two cases the power stations cover the *full* requirement. These factors cause the differences to be noted in the breakdown.

Table 18. *Examples of the investment costs for integrated iron and steelworks*

	Example 1	Example 2		Example 3	
Crude steel capacity in t/y Total investment costs in DM and $ per annual ton capacity	300,000 1,104. — DM 276 $	800,000 1,236. — DM 309 $		2,400,000 885. — DM 221 $	
	%	%	%ᵃ	%	%ᵃ
Coking plant	—	6.5	—	6.6	—
Blast-furnace plant including sintering plant	16.2	7.7	8.2	10.4	11.1
Oxygen steelworks	11.9	10.7	11.5	12.0	13.0
Hot rolling mills	38.2	30.5	32.6	14.8	16.0
Cold rolling mills	—	—	—	6.3	6.7
Rolling mills, total	38.2	30.5	32.6	21.1	22.7
Power station	3.2	5.6	6.0	6.5	6.9
Power and energy supplies (electric power, gas and water distribution)	8.8	12.3	13.2	13.4	14.3
Works transport and handling facilities, including harbour and raw materials stores	6.2	12.8	13.7	13.5	14.4
Workshops and other general plant	12.4	6.8	7.3	6.5	7.0
Auxiliary and ancillary plant: Total	27.4	31.9	34.2	32.4	35.7
Miscellaneous	3.1	7.1	7.5	10.0	10.6
Totals:	100	100	100	100	100

ᵃ The same values, calculated *without* coking plant.

Table 19 gives a different breakdown of the costs shown in Table 18, i.e. for production shops, etc., and also for the mechanical facilities (mostly fob supplies), erection, and building work (foundations, etc.).

Table 19. *Summary of investment costs for integrated iron and steelworks*
(see also Table 18)

	Example 1		Example 2		Example 3	
	300,000 t		800,000 t		2,400,000 t	
	%		%		%	
Breakdown of departments:						
Production shops	66.3		55.4		50.1	
Power station	3.2		5.6		6.5	
Auxiliary, ancillary, and general plant	30.5		39.0		43.4	
Total:	100.0		100.0		100.0	
Breakdown into mechanical and electrical equipment (= fob supplies), building work and erection costs:						
Mechanical and electrical equipment (fob supplies)	66.0	100.0	51.5	100.0	52.4	100.0
Building work (buildings, foundations, etc.)	24.9	37.8	38.1	74.1	37.4	71.4
Erection costs	9.1	13.8	10.5	21.4	10.2	19.5
Total:	100.0	151.6	100.0	195.5	100.0	190.9

The following estimated costs for integrated iron and steelworks
are based on the costs given in these two tables, and on empirical
values obtained during the planning of iron and steelworks in the
recent past.   These figures include the safety margins normally
headed "contingency allowances". Case a) covers an iron and steelworks with its own coking plant and a power station for 100%

Guide values for the investment costs of integrated iron and steelworks

	a) With coking plant, and power station rated at 100% %	b) Without coking plant, but with a power station rated at 25% %
Coking plant	6.5	—
Blast furnace plant[a]	10.0	12.0
Oxygen steelworks	12.0	14.0
Rolling mills	32.0	36.0
Power station	6.5	4.0
General plant	33.0	34.0

[a] Including sintering plant, etc.

coverage, Case b) an iron and steelworks *without* coking plant and with a power station designed to cover 25% of the demand.

The comparison shows that in each case, one-third of the investment costs is accounted for by the rolling mills and general plant, and that the blast-furnace plant and steelworks balance. Naturally, these guide values are based on estimates; the figures quoted could be quite different if other steelmaking processes are used or if the rolling mills are designed for the production of semi-finished products only, for example.

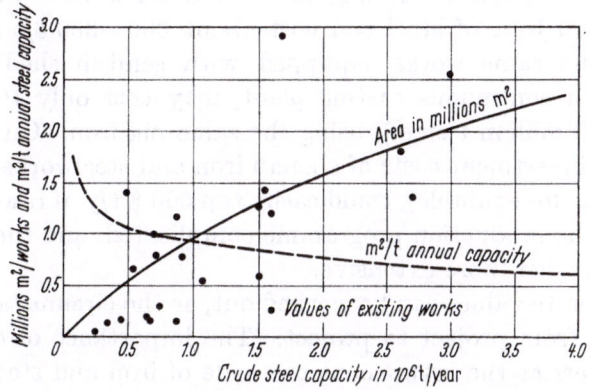

Fig 38. Space requirements of integrated iron and steelworks.

Fig. 38 gives an idea of the ground space required for an integrated iron and steelworks. The curves for total required area in millions of m² and for m² per ton annual capacity were plotted on the basis of planning work undertaken for some 25 iron and steelworks projects in recent years.

X. General Remarks

Even though prices have in general risen quite severely over the past fifteen years or so, the costs of individual items of equipment for metallurgical works, and in some measure of complete iron and steelworks, not only remained constant, but actually dropped during the same period.

This is a direct result of technical developments made during this time; the equipment is of the same size as in earlier days, but efficiency and output have been greatly increased.

Raw materials play a very important role when the costs of individual items of plant and equipment are being estimated. For example, blast furnace investment costs per ton of pig iron produced are affected by the Fe content of the ore charged; an ore with 40% Fe will produce much higher costs than an ore with 60% Fe. Steelworks investment costs are affected by the pig iron quality, i.e. it may be necessary to use the Kaldo or LDAC process instead of the normal top-blowing oxygen process if P contents are excessively high. The problem of establishing generally valid guide values for rolling mills has already been outlined.

For instance, an integrated iron and steelworks with a normal pig iron basis, normal rolling mills, and an annual capacity of 2—3 million tons of steel can cost about 300—350 $/t steel p.a., whereas the same works, equipped with semi-finished products mills and a continuous casting plant, may cost only 100 $/t p.a. (capacity 3 million t steel), using the same pig iron. On the other hand, the investment costs of a small iron and steelworks (capacity 300,000 t/y, for example), could easily top 500 $ t/y, if raw materials are poor, the production programme complicated, and the finishing processes rather more extensive.

No concrete values can be worked out, as the circumstances vary so greatly from project to project. The importance of entrusting such matters as the investment cost side of iron and steelworks to experienced firms or consultants cannot be over-emphasised.

E. Profitability

The yardstick of profitability is naturally the amount of net profit made each year; in joint-stock companies, for example, it is the amount distributed in dividends. A distinction is made between gross and net profit. Gross profit is the sum left when the costs of raw materials, labour and so on are deducted from the turnover. Interest, taxes, and similar sums are then deducted from the gross profit to leave the net profit.

The term "cash-flow" is used to describe a rapid method of measuring profitability; this method is used mainly in the USA. The cash-flow figure is obtained from the relationship between profit and working capital, i.e. the investment costs plus operating capital.

The cash-flow index is calculated by establishing the turnover and costs for a specific period of time, deducting the costs from the turnover, and dividing the result by the working capital.

Generally, the cash-flow index is expressed as a percentage, but it can also be expressed as a definite sum.

The formula used is:

$$\frac{\text{Profit}}{\text{working capital}} = \text{cash-flow}$$

$$\text{Example:} \frac{8{,}000{,}000\ \$}{42{,}500{,}000\ \$} \times 100 = 18.8\ \%$$

Fig. 39 shows the cash-flow figures for varying degrees of profit. The values given in the diagram do not relate to any particular works, and are used only as an example.

The cash-flow figure established here (18.8% = 8.0 million $) includes depreciation, interest, and profit after deduction of taxes.

Profitability calculations also cover the establishment of the break-even point. This point is reached when the turnover (sales returns) equals the production costs, or in other words when the production rate is such that neither profit nor loss is made.

In order to establish the break-even point, the production costs must first be divided into fixed and variable elements. When these

are known, future alterations in total costs or costs per production
unit on the production rate increasing or dropping can be determined.

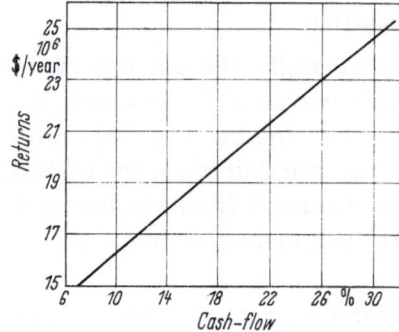

Fig. 39. Cash-flow diagram.

Basis: Investment costs 40,0 millions $;
operating assets 2,5 millions $; sales
returns 20,0 millions $; costs 12.0 $;
profit 8.0 $; cash-flow 18.8%.

Fig. 40 shows the profitability situation as a function of the
production rate.

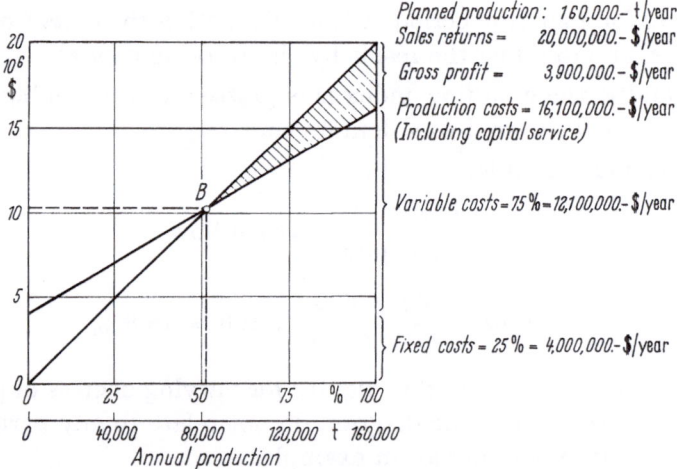

Fig. 40. The break-even point is reached at 82,400 t = 51.5% of the planned
annual production.

In this example, the fixed costs were estimated at 25% and the
variable costs at 75%.

The annual production rates are plotted along the abscissa, and
the sales returns and production costs (divided into fixed and variable costs) are plotted along the ordinate.

In the example shown, the break-even point is reached when 51.5% of the planned production figure is attained (82,400 t produced, 160,000 t planned).

The degrees of loss or profit made at varying production rates can also be taken from the diagram.

F. Annexures

Conversion Tables and Factors

Length	1 mm = 0.03937 inch = 39.37 mils
	1 cm = 0.3937 inch
	1 m = 39.37 inches = 3.2808 feet = 1.0936 yard
	1 km = 1093.6 yards = 3280.8 feet = 0.62137 mile
Area, Surface	1 cm² = 1.55 square inches
	1 m² = 10.7639 square feet = 1.19599 square yards
Volume	1 cm³ = 0.061 cubic inch
	1 dm³ (Liter) = 1.057 US liquid quart = 1.7598 pints (Imp.)
	1 m³ = 35.3166 cubic feet = 1.308 cubic yard
Weight	1 g = 0.035 ounce
	1 kg = 2.2046 lbs
	1 t = 1000 kg = 0.9842 long ton (Brit.) or gross ton (US)
	= 1.1023 short ton (Brit.) or net ton (US)
Pressure	1 mm WS = 0.03937 inch water
	1 at = 1 kg/cm² = 14.2234 lbs/square inch
Temperature	1 °C = (9/5 + 32) °F
	n °C = (n 9/5 + 32) °F
Calorific value	1 kcal = 3.968 B.T.U.
	1 kcal/kg = 1.80 B.T.U. per lb.
	1 kcal/m³ = 0.1124 B.T.U. per cubic foot
Energy (Power)	1 PS = 0.986 HP
	1 kW = 1.341 HP = 0.948 B.T.U. per sec.
	1 kWh = 3,413 B.T.U.
Work	1 mt = 7,233 ft. lbs.

Centigrade to Fahrenheit

°C	°F	°C	°F	°C	°F
− 20	− 4	120	248	1,200	2,192
− 15	+ 5	140	284	1,300	2,372
− 10	+ 14	160	320	1,400	2,552
− 5	+ 23	180	356	1,500	2,732
0	+ 32	200	392	1,600	2,912
5	41	300	572	1,700	3,092
10	50	400	752	1,800	3,272
20	68	500	932	1,900	3,452
30	86	600	1,112	2,000	3,632
40	104	700	1,292	2,100	3,812
50	122	800	1,472	2,200	3,992
60	140	900	1,652	2,300	4,172
70	158	1,000	1,832	2,400	4,352
80	176	1,100	2,012	2,500	4,532
90	194				
100	212				

Millimetres to inches

mm	inches	mm	inches	mm	inches
1	0.039	45	1.772	350	13.779
2	0.079	50	1.968	400	15.748
3	0.118	55	2.165	450	17.716
4	0.157	60	2.362	500	19.685
5	0.197	65	2.559	550	21.653
6	0.236	70	2.756	600	23.622
7	0.275	75	2.953	650	25.591
8	0.315	80	3.150	700	27.559
9	0.354	85	3.346	750	29.528
10	0.394	90	3.543	800	31.496
15	0.590	95	3.740	850	33.465
20	0.787	100	3.937	900	35.433
25	0.984	150	5.905	950	37.402
30	1.181	200	7.874	1.000	39.370
35	1.378	250	9.984		
40	1.575	300	11.810		

Kilogram Calories to B.T.U.

kcal	B.T.U.	kcal	B.T.U.
1	3.97	1,000	3,968
5	19.80	1,200	4,762
10	39.7	1,400	5,556
20	79.4	1,600	6,349
50	198.4	1,800	7,143
100	396.8	2,000	7,937
200	794	3,000	11,905
300	1,190	4,000	15,873
400	1,587	5,000	19,841
500	1,984	6,000	23,810
600	2,381	7,000	27,778
700	2,778	8,000	31,746
800	3,175	9,000	35,715
900	3,571	10,000	39,683

Kilograms to pounds

kg	lbs	kg	lbs	t (metr.)	short/net ton	long/gross ton
1	2.205	55	121.2	1	1.102	0.984
2	4.409	60	132.3	2	2.204	1.968
3	6.614	65	143.3	3	3.306	2.952
4	8.818	70	154.3	4	4.408	3.936
5	11.023	75	165.3	5	5.510	4.920
6	13.228	80	176.3	6	6.612	5.904
7	15.432	85	187.4	7	7.714	6.888
8	17.637	90	198.4	8	8.816	7.872
9	19.842	95	209.4	9	9.918	8.856
10	22.046	100	220	10	11.02	9.84
15	33.1	200	441	50	55	49
20	44.1	300	661	100	110	98
25	55.1	400	882	500	551	492
30	66.1	500	1,102	1,000	1,102	984
35	77.2	600	1,323			
40	88.2	700	1,543			
45	99.2	800	1,764			
50	110.2	900	1,984			
		1,000	2,205			

Cubic metres to cubic feet

m³	cu. ft	m³	cu. ft
1	35.315	65	2,295
2	70.630	70	2,472
3	105.944	75	2,649
4	141.259	80	2,825
5	176.574	85	3,002
6	211.889	90	3,178
7	247.203	95	3,355
8	282.518	100	3,531
9	317.832	200	7,063
10	353.148	300	10,594
15	528.7	400	14,126
20	706.3	500	17,657
25	882.8	600	21,190
30	1,059	700	24,720
35	1,236	800	28,252
40	1,413	900	21,784
45	1,589	1,000	35,315
50	1,766	5,000	176,570
55	1,942	10,000	353,150
60	2,119		

Square metres to square feet/yards

m²	sq.ft	sq.yd	m²	sq.ft	sq.yd
1	10.76	1.196	55	592.0	65.78
2	21.53	2.392	60	645.8	71.76
3	32.29	3.588	65	699.7	77.70
4	43.06	4.784	70	753.5	83.72
5	53.82	5.980	75	807.3	89.70
6	64.58	7.176	80	861.1	95.68
7	75.35	8.372	85	914.9	101.7
8	86.11	9.568	90	968.7	107.6
9	96.87	10.764	95	1,022	113.6
10	107.6	11.96	100	1,076	119.6
15	161.5	17.94	200	2,153	239
20	215.3	23.92	300	3,229	359
25	269.1	29.90	400	4,306	478
30	322.9	35.88	500	5,382	598
35	376.7	41.86	600	6,458	718
40	430.6	47.84	700	7 535	837
45	484.4	53.82	800	8,611	957
50	538.2	59.80	900	9,687	1,076
			1,000	10,764	1,196

Abbreviations

The following abbreviations are used in this book:

C	Centigrade		m	metre
c/s	cycles		m^2	square metre
d	day (24 h)		m^3	cubic metre
Dpf	Deutschpfennig = 0.01 DM		mm	millimetre
DIN	German Industrial Standards		mm^2	square millimetre
g	gram		mt	metre ton
h	hour		MW	megawatt
kcal	kilogram calorie		Nm^3	standard cubic metre
				(0 °C, 760 mm mercury column)
kg	kilogram		p.i.	pig iron
km	kilometre		PS	metric HP
kW	kilowatt		s	second
kWh	kilowatt hour		t	metric ton
l	litre		y	year

The oblique stroke (/) is used to denote "per", for example:

Nm^3/h standard cubic metres per hour;

t/y   metric tons per year.

The remaining abbreviations used in this book are common to all English-speaking countries and are therefore not listed above.

Bibliography

A. General Conditions

The Making, Shaping and Treating of Steel, Publishers: United States Steel Corporation. 7th Edition, Pittsburgh, 1957, XXIII, 1048 pages, 4°.

Watkins Cyclopedia of the Steel Industry, Publishers: Steel Publications, Inc., Pittsburgh.

B. I. Fuels

Fuel Efficiency Bulletin, Publishers: Ministry of Fuel and Power (London). Committee on the Efficient Use of Fuel, No. 44. The Operation of Gas Producers. 1946, 32 pages.

BASHFORTH, G. REGINALD: The Manufacture of Iron and Steel. Vol. 3, Steelworks Fuels, Furnaces, Refractories and Instruments. 1960, VIII, 246 pages.

ETHERINGTON, H., and G. ETHERINGTON: Modern Furnace Technology. 3rd Edition, Rev., London: Griffin 1961, XII, 246 pages.

Technical Data on Fuel. Ed. by H. M. SPIERS, 6th Edition, compl. revised. Publishers: The British National Committee, World Power Conference, London 1961, XI, 360 pages, 4°.

Chemical Engineering in the Coal Industry. An International Conference Organized by the National Coal Board, Great Britain, and Held at its Coal Research Establishment at Cheltenham, England, in June 1956. Ed. by FORBES W. SHAPLEY, London/New York/Paris: Pergamon Press 1956, 141 pages, 8°.

Construction and Operation of Metallurgical Coke Ovens. Publishers: Unites Nations, Economic Commission for Europe. New York 1965. 76 pages, 4°.

TAGGART, ARTHUR F.: Elements of Ore Dressing, New York, Wiley, London: Chapman & Hall 1951, XVII, 595 pages, 8°.

Recent Developments in Mineral Dressing. A Symposium arranged by the Institution of Mining and Metallurgy, held on 23rd—25th September, 1952, at the Huxley Building, Imperial College of Science and Technology. Publisher: Institution of Mining and Metallurgy. London 1953, XXVIII, 766 pages, 8°.

B. II. Metallurgical Processes

a) Ore Preparation and Ore Beneficiation
b) Ore Reduction

Progress in Mineral Dressing. Transactions of the International Mineral Dressing Congress Stockholm 1957. Sponsored and ed. by Svenska Gruvföreningen and Jernkontoret. Stockholm: Almquvist & Wiksell 1958, 754 pages.

Agglomeration. Ed. by WILLIAM A. KNEPPER. Based on an International Symposium held in Philadelphia, PA., April 12—14, 1961. Sponsored by American Institute of Mining, Metallurgical and Petroleum Engineers, in cooperation with the Society of Mining Engineers. The Metallurgical Society and the Society of Petroleum Engineers. New York/London: Interscience Publishers 1962. XV, 1109 pages.

Ore Mining and Materials Handling. Papers and Discussions from the 47th Iron
and Steel Engineers Group Meeting held, with the collaboration of the Chambre
Syndicale de la Sidérurgie Française and the Société Française de Métallurgie,
in Metz, France, on 10—14 June, 1963. 85 pages.

Electric Furnaces. Ed. by C. A. OTTO, London: Newnes 1958, VII, 240 pages, 8°.

FINE, M. M., J. P. HANSEN and NORWOOD B. MELCHER: Prereduced Iron Ore
Pellets: A New Blast Furnace Raw Material. Publishers: U.S. Department of the
Interior, Bureau of Mines, Washington, 1962, 19 pages, 4°. Report of Investiga-
tions 6152.

Iron Ore Reduction. Proceedings of a Symposium of the Electrothermics and
Metallurgy Division of the Electrochemical Society, held in Chicago, 3—5 May,
1960. Sponsored by the Electrochemical Society, New York; Ed. by R. R. ROGERS,
Oxford/London/New York/Paris: Pergamon Press 1962. XV, 359 pages, 4°.

c) Steelworks

JACKSON, A.: Steelmaking for Steelmakers. Publishers: United Steel Companies
Ltd., Sheffield 1959. XI, 265 pages, 8°.

BASHFORTH, G. REGINALD: The Manufacture of Iron and Steel. 3rd Ed., Vol. 2,
Steel Production 1964. 479 pages.

Comparison of Steelmaking Processes. Publishers: United Nations, Economic
Commission for Europe. New York 1962, VIII, 83 pages, 4°.

European Progress in Pneumatic Steelmaking Methods. COHEUR, Pierre; Metal
Progr. 81 (1962) No. 1, p. 100/03 and 133/34.

Basic Open Hearth Steelmaking; by the Physical Chemistry of Steelmaking
Committee, Iron and Steel Division, AIME. Ed. by W. O. PHILBROOK and M. B.
BEVER in collaboration with H. B. EMERICK and B. M. LARSEN. 2nd Ed.,
completely revised and enlarged. New York: American Institute of Mining and
Metallurgical Engineers, 1951. XIX, 940 pages, 8°.

The All-Basic Open-Hearth Furnace. Being also the report of the 36th Steelmaking
Conference, held at Ashorne Hill-Leamington Spa, on 2—3 May, 1951. Prepared
by the All-Basic Open-Hearth Furnace Sub-Committee, British Ceramic Research
Association and British Iron and Steel Research Association. 1952, VI, 92 pages.

LYCHAGIN, A. S.: Open Hearth Furnace Design. (Translated from the Russian.)
Published in association with Department of Scientific and Industrial Research.
London: Butterworth 1962, IX, 246 pages, 4°.

Basic Open Hearth Steelmaking. With supplement on oxygen in steelmaking. By
the Physical Chemistry of Steelmaking Committee, Iron and Steel Division, The
Metallurgical Society of AIME. 3rd ed. Ed. by GERHARD DERGE. Publ. by The
American Institute of Mining, Metallurgical and Petroleum Engineers. New York
1964, XXII, 1007 pages, 8°.

WISSMANN, CONRAD C.: Acid Electric Furnace Steelmaking Practice. Cleveland,
Ohio: American Society for Metals 1947. 84 pages, 8°.

PASCHKIS, V. and JOHN PERSSON: Industrial Electric Furnaces and Appliances.
2nd ed., New York: Interscience Publishers 1960. XVI, 607 pages, 8°.

Electric Furnace Steelmaking. Vol. 1. 2. Sponsored by: Physical Chemistry of Steel-making Committee, Iron and Steel Division, American Institute of Mining, Metallurgical, and Petroleum Engineers. Ed. by CLARENCE E. SIMS, New York/ London: Interscience Publishers. 8°. Vol. 1. Design, Operation, and Practice. 1962, XV, 404 pages. Vol. 2. Theory and Fundamentals. 1963, XV, 471 pages.

CHARLES, J. A., W. J. B. CHATER and J. L. HARRISON: Oxygen in Iron and Steel Making. London: Butterworth 1956, XII, 309 pages, 8°.

COHEUR, P.: Pneumatic Processes for Converting. J. Metals 12, 1960, No. 7, pp. 545/47.

TRENTINI, B., P. VAYASSIERE and C. ROEDERER: OLP Steelmaking-Recent Studies and Industrial Progress. J. Metals 13, 1961, No. 6, pp. 418/21.

JONES, D. J., A. E. PARSONS and N. MORRIS: LD and LDAC Operating Experience in Britain. J. Metals 15, 1963, No. 8, pp. 577/80.

COLMANT, R. H.: Kaldo Moves Forward. J. Metals 14, 1962, No. 7, pp. 505/08.

d) Finishing Process

KOROTKOV, K. P., H. P. MAYOROV, A. A. SKVORTSOV and A. D. AKIMENKO: The Continuous Casting of Steel in Commercial Use. Transl. from the Russian by V. ALFORD. Ed. by H. T. PROTHEROE, Oxford: Pergamon Press 1960. 171 pages, 8°.

BOICHENKO, M. C.: Continuous Casting of Steel. Ed. by G. FENTON. Published in association with the Department of Scientific and Industrial Research, London: Butterworth 1961, XI, 218 pages, 8°.

Continuous Casting of Steel. Report of the proceedings of the Autumn general meeting of the Iron and Steel Institute, held on 25 and 26 Nov., 1964, at London. 1965. 195 pages.

Mechanical Working of Steel. I. Proceedings of the 5th technical conference, sponsored by the Mechanical Working and Steel Processing Committee of the Iron and Steel Division, The Metallurgical Society and the Pittsburgh Section, American Institute of Mining, Metallurgical and Petroleum Engineers, Pittsburgh, Pa., Jan. 15—16, 1963. Ed. by PHILLIP H. SMITH, New York: Gordon & Breach 1964, VIII, 417 pages, 8°.

Recent Progress in Metal Working. Lectures delivered at the Institution of Metallurgists refresher course, Oct. 1963. Publ. for the Institution of Metallurgists, New York: American Elsevier Publ.; London: Iliffe Books 1964. 136 pages, 8°.

BEYNON, ROSS E.: Roll Design and Mill Layout. Publ. by the Association of Iron and Steel Engineers. Pittsburgh, 1956, 178 pages, 4°.

POLUCHIN, P. I., N. M. FEDOSOV, A. A. KOROLEV and Ju. M. MATVEEV: Rolling Mill Practice. Transl. from the Russian by NICHOLAS WEINSTEIN. Moscow: Peace Publ. 1960, 510 pages, 4°.

LARKE, EUSTACE C.: The Rolling of Strip, Sheet and Plate; 2nd ed. London: Chapman & Hall 1963, XIV, 449 pages, 8°.

B. III. Ancillary and Auxiliary Departments

MACQUEEN, A. J. F.: Factors in the Design of a Steel Plant Power System. Iron Steel Eng. 35, 1958, No. 10, pp. 92/100.

STEWART JR., DONALD: The Power Distribution System of an Integrated Steel Plant. Iron Steel Eng. 38, 1961, No. 12, pp. 150/55.

ZITTEL, T. O. and Z. A. BIENIULIS: Operational Aspects of Power Provision, Distribution and Utilisation in an Integrated Steel Plant. Iron Steel Eng. 40, 1963, No. 10, pp. 129/36.

Manual on Industrial Water. 4th printing with new and revised methods. Publishers: American Society for Testing Materials. Philadelphia 1957, X, 490 pages, 8°.

Water Economy in Iron and Steelworks. Publ. by the European Productivity Agency of the Organisation for European Economic Co-operation. Paris 1958. 70 pages, 4°.

Glossary of Terms for Producers and Users of Iron Castings. Publishers: The International Nickel Company. New York (about 1955), 36 pages.

Steel Castings Handbook. 3rd ed. Ed.: CHARLES W. BRIGGS. Publishers: Steel Founders' Society of America. Cleveland, Ohio, 1960, VIII, 670 pages, 8°.

JOSEPHSON, G. W., F. SILLERS, JR., and D. G. RUNNER: Iron Blast-Furnace Slag. Production, Processing, Properties, and Uses. Publishers: United States Department of the Interior. Bureau of Mines. Washington, 1949, XIII, 304 pages, 8°.

NIJHAWAN, B. R.: Research and Development Work on the Utilisation of Metallurgical Wastes at the National Metallurgical Laboratory. NML, Techn, J. 6, 1964, No. 1, pp. 50/59.

TRINKS, W.: Calcination or Burning of Limestone. Ind. Heat. 29, 1962, No. 3, pp. 442/46 and 48.

MIDDLEMAS, J. W.: Calcination of Limestone and the Development of Oil-fired Kilns. J. Inst. Fuel 36, 1963, No. 269, pp. 244/53.

MOXLEY, T. R.: Steel Plant Maintenance Shops. Iron Steel Eng. 18, 1941, No. 1, pp. 75/82.

PEARSON, W. J.: Ideal Maintenance Machine Shop Facilities. Iron Steel Eng. 33, 1956, No. 9, pp. 75/80.

Maintenance Engineering Handbook. Publishers: L. C. Morrow. New York, McGraw-Hill 1957, 8°.

WATSON, HAROLD E.: Electrical Maintenance at Pueblo Plant of the Colorado Fuel & Iron Corp. Iron Steel Eng. 34, 1957, No. 6, pp. 104/10.

CORDER, G. G.: Organising Maintenance. Publishers: British Institute of Management. London 196, VIII, 44 pages, 8°.

STEWART, H. V. M.: Guide to Efficient Maintenance Management. London: Business Publications 1963, X, 157 pages, 8°.

BASHFORTH, G. REGINALD: Mechanical Handling in the Iron and Steel Industry. Brit. Steelmaker 16, 1950, No. 6, pp. 281/92.

KAIDANOWSKY, S. P.: Guide to Materials, Standards and Specifications. Part 2: Iron and Steels. Mater. Design Engineering 47, 1958, No. 4, pp. 110/14.

National Steel Specifications. Report of the proceedings of the Annual General Meeting of The Iron and Steel Institute held at The Institute of Civil Engineers, London 1964. 215 pgges.

British Steel Standards. Engineer 218, 1964, No. 5658, pp. 10/11.

MONTORO, J. C.: Use of Models for Layout and Planning. Iron Steel Eng. 31, 1954, No. 9, pp. 154/60.

MALLICK, RANDOLPH W., ARMAND T. GAUDREAU: Plant Layout, Planning, and Practice. New York: Wiley & Sons: London; Chapman & Hall (1951). XII, 391 pages, 8°.

Internal Transport in Iron and Steel Works. Publ. by the European Productivity Agency of the Organisation for European Economic Co-operation. Paris (about 1956), 216 pages, 4°.

BASHFORTH, G.: Some Aspects of Melting Shop Management. Brit. Steelmaker 16, 1950, No. 7, pp. 368/74, No. 8, pp. 408/20, No. 9, pp. 450/55.

YOUNG, A. J.: An Introduction to Process Control System Design. London, Longmans, Green, 1955, XVII, 379 pages, 8°.

Statistical Quality Control. Publ. by the European Productivity Agency of the Organisation for European Economic Co-operation. Paris, 1956, 89 pages, 8°.

ALLAN, DOUGLAS H. W.: Statistical Quality Control. An introduction for management. New York: Reinhold, London: Chapman & Hall 1959, 129 pages, 8°.

LAWRENCE, A. E.: Production Planning for an Integrated Steel Plant. Iron Steel Eng. 37, 1960, No. 5, pp. 133/39.

C. The Planning of Iron and Steelworks
D. Investment Costs
E. Profitability

ALFORT, L. P., and JOHN R. BANS: Production Handbook, New York, Ronald Press Company 1948, IX, 1676 pages, 8°.

BODDINGTON, H.: An Architect's Approach to the Design and Layout of Modern Iron and Steel Works. Iron Coal Trades Rev. 168, 1954, No. 4474, pp. 83, No. 4475, pp. 131/38.

MOSSO, A. J.: Cost Estimating — Metal Working Plants. Iron Steel Eng. 36, 1956, No. 9, pp. 87/99.

BRISBY, M. D. J., P. M. WORTHINGTON and R. J. ANDERSON: Economics of Process Selection in the Iron and Steel Industry. J. Iron Steel Ind. 202, 1964, No. 9, pp. 721/34, cf. Iron & Steel 37, 1964, No. 6, pp. 255/61.

The Steel Development Plan in Action No. 18; The Steel Company of Wales Limited. Iron Coal Trades Rev. 159, 1949, No. 4255, pp. 100/10 and 119.

KNIGHT HAROLD A.: Plan for East Coast Steel Plant. J. Metals 1, 1949, No. 2, pp. 6/9.

MARDON, H. H., and J. S. TERRINGTON: The Layout of Integrated Iron and Steelworks. J. Iron Steel Inst. 161, 1949, No. 4, pp. 327/59, cf. Iron Coal Trades Rev. 159, 1949, No. 4255A, pp. 37/38.

ATKINS, W. S. A.: New Steelworks. Port Talbot — Planning and Design. Engineer, London 189, 1950, No. 4915, pp. 417/19.

McQUAID, H. W.: Small Steel Mills for Local Markets. Iron Age 165, 1950, No. 14, pp. 90/94.

Britain's Newest Steelworks Opened. Steel Company of Wales Project Materialises. Metallurgia, Manchester, 44, 1951; No. 262, pp. 63068.

The Works of the Steel Company of Wales I/VII. Engineer, London, 192, 1951;
No. 4997, pp. 554/56, No. 4998, pp. 583/85, No. 4999, pp. 620/21, No. 5000,
pp. 649/52, No. 5001, pp. 690/91, No. 5002, pp. 720/22, No. 5003, pp. 756/59.

KESTERTON, A. J.: Some Starting and Operating Experiences at Abbey Melting
Shop. J. Iron Steel Inst. 179, 1955; No. 1, pp. 46/57.

Construction at Spencer Works. Engineer 214, 1962; No. 5563, pp. 405/10.

CARTWRIGHT, W. F.: The Steel Company of Wales Limited — Its Development to
Date. Iron Steel Engineer. 40, 1963; No. 5, pp. 69/96.

Index